ENTOMOLOGY

An Illustrated Textbook

NEW INDIA PUBLISHING AGENCY

New Delhi – 110 034

About the Author

Dr. P.K. Sehgal (b-1980) completed his M.Sc Zoology/Entomology in 2004 and done Ph.D Zoology/Entomology in 2008. He has 18 instant notes and 13 books, 19 Research Papers to his credit. At Presently he is working as Assistant Professor in the Department of Zoology/ Entomology at Dolphin (P.G.) Institute of Biomedical and Natural Sciences, Dehradun, Uttarakhand, India. The books run like wild fair from city to city, state to state and crossed the frontier of the country. This book is helpfull for B.Sc Agricultural student for fundamental knowledge of Entomology. Several books for undergraduate and postgraduate students and competitive examination of state and National level followed in quick succession. The Language is simple, explanation are clear and presentation is very systematic.

ENTOMOLOGY

An Illustrated Textbook

P.K. SEHGAL (M.Sc. J.R.F., S.R.F, Ph.D)

Assistant Professor
Department of Zoology/Entomology
Dolphin (PG) Institute of Biomedical and Natural Sciences
Dehradun, Uttarakhand (India)

NEW INDIA PUBLISHING AGENCY
101, Vikas Surya Plaza, CU Block, LSC Market
Pitam Pura, New Delhi 110 034, India
Phone: + 91 (11)27 34 17 17 Fax: + 91(11) 27 34 16 16
Email: info@nipabooks.com
Web: www.nipabooks.com

Feedback at feedbacks@nipabooks.com

ISBN: 978-93-95763-95-0

Composed, and Designed by NIPA

Dedicated to
our Respected Parents
and
All my Teachers

Preface

It gives us immense pleasure to put before you. **"Entomology: An Illustrated Texbook"** The book has been written in simple and lucid language keeping in view all the grades of students. While writing this book, a number of text books, journals, encyclopedias have been consulted. We therefore express our gratitude to the authors, eminent scientists and the publishers of those books who have put their efforts in producing such scientific literature.

It is hoped that this textbook will serve the purpose of students at under-graduate and post-graduate levels of different Indian universities. The keen interest and the sincere devotion of the publisher of this book needs the special appreciation. We are very much thankful to all those who have directly or indirectly helped us in bringing this book.

P.K. Sehgal

Acknowledgements

We express our gratitude to all those who made significants contributions in the improvement of manuscript of this book through their critical analysis, invaluable suggestions and encouraging opinions. Some important dignitaries among them are:

Dr. S.C. Dhiman (EX-Reader, Dept of Zoology, M.S (PG) College, Saharanpur (UP).

Dr. S. Kumar (Associate Professor Dept of Zoology- Raza College, Rampur, Dist UP).

Dr. M.C. Sharma (Ex-Reader and HOD, convener R.D.C. Ph.D C.C.S. University Meerut).

Dr. Shailja Pant (Principal, Dolphin (PG) Institute of Biomedical and Natural Sciences, Dehradun Uttarakhand).

Dr. Arun Kumar (Director, Dolphin (PG) Institute of Biomedical and Natural Sciences, Dehradun Uttarakhand).

Shri Arvind Gupta (Chairman, Dolphin (PG) Institute of Biomedical and Natural Sciences, Dehradun Uttarakhand).

Dr. Y.K. Mathur (HOD) of Entomology (C.S.A.U.A.T.) Kanpur UP.

Dr. Rajendra Singh, Reader, Department of Zoology, D.D.U. Gorakhpur University, Gorakhpur.

Smt. Pratibha Rani and My small child, Anshyia Sehgal, Ansh Sehgal, Gracy Shri Sehgal.

About Entomology

SYLLABUS FOR B.Sc. AGRICULTURAL ENTOMOLOGY

The scope of Entomology, brief history of entomology in India, insects as Arthropods and its relationship with phylum Annelida and other classes of Arthropoda, origin in insects major points related to dominance of insects in Animal Kingdom. External morphology and anatomy of grass hopper; body segmentation, integument, thorax and abdomen, antennae, legs and wings and their modifications, generalized mouth parts and their modifications, Alimentary, Circulatory, Excretory, Respiratory, Reproductive and nervous systems, major sensory organs like simple and compound eyes chemoreceptors, endocrine glands; basic embryology and post embryonic development, basic groups of present day insects with special emphasis to orders and families of agricultural importance like Orthoptera; Tetigonidae, Gryllidae, Gryllotalpidae, Acrididae, Dictyoptera; Mantidae, Blattidae; Isoptera; Hemiptera; Pentatomidae; Coreidae; Cimicidae, Cicadellidae, Delphacidae, Lophophidae, Aleurodidae; Aphididae; Coccidae; Thysanoptera, Coleoptera. Carabidae, Meloidae, Coccinellidae, Bruchidae, Chrysomelidae, Curculionidae, Cerambycidae; Diptera; Culicidae Cephritidae, Agromyzidae, Muscidae; Lepidoptera, Pleridae; Papilionidae, Hespirlidae, Sphingidae, Noctuidae, Artilidae, Pyralidae, Saturnidae, Bombycidae; Hymenoptera. Tenthredinidae, Braconidae, Chalcididae, Trichogrammatidae. How insects become pest economic importance of insects, classification of pests, principles and methods of pest control, viz, physical mechanical, cultural, legal, genetical chemical. Biological, principles and methods of insecticidal applications, Apiculture, Sericulture and lac cultivation with special reference to equipment used insect pests and diseases, production and marketing.

Practicals

Collection killing, planing and mounting of insects, study of different classes of phylum Arthropoda, external morphology of grasshopper, typical mouth parts and their modification of antennae, legs, wings and their coupling apparatus, structure of alimentary canal and nervous system, Michaela, reproductive and other systems in insects, post embryonic development in insects and basic of insects classification. Basic groups of present day insects with special reference to orders and families of agricultural importance.

AGRICULTURAL ENTOMOLOGY

Definition of Entomology: Entomology, as the name suggests, is the study of insects in relation to Agriculture and forest product. The term entomology has been derived from the Greek words entomon = insect and, logos = knowledge, thus in broad sense it may be defined as a branch of zoology which deals with the insects. It is a specialized field of entomology with the main domain of zoology and deals specifically with problems of forestry and the insects which feed on them. It differs from other branches of entomology.

1. Agricultural Entomology
2. Medical Entomology
3. Veterinary Entomology
4. Forest Entomology

Large number of insects attack valuable tree and are called the forest insect pests. They cause heavy loss every year. It is a difficult to make an assessment of the revenue loses caused by the forest insect pests in a vast and diverse country like India. The identification and control of the most common or serious insect pests that affect plantation and trees in the forests are discussed.

SCOPE OF ENTOMOLOGY

Encyclopedia of Entomology brings together the expertise of distinguished entomologist to provide a world wide overview of insects and their close relatives. Combining the basic science of an introductory text with accurate, comprehensive detail, the encyclopedia is a reliable first source of references for student and working professionals. Coverage includes insect classification, behavior, ecology, genetics and evolution, physiology and management and references to relevant literature. All the major arthropods groups are add reseed, along with many important family and species. The encyclopedia places special emphasis on insect relationships with peoples, medical entomology. Biological control and insect Pathology. This important work also presents boo-graphical sketches of hundreds of entomologist who have made important contributions to the discipline since its origin. (Springer 2015).

GENERAL CHARACTERS OF PHYLUM ARTHROPODA

1. Arthropods are triploblastic, bilaterally symmetrical, metamerically segmented animals.

2. Body is covered with thick chitinous cuticle forming an exoskeleton.
3. Body segments usually bear paired lateral and jointed appendages.
4. Body cavity is haemocoel. The true coelom is reduced to the spaces of the genital and excretory organs.
5. Digestive tract is complete; mouth and anus lie at apposite end of the body.
6. Musculature is not continuous but comprises separate striped muscles.
7. Circulatory system is open with dorsal heart and arteries but without capillaries.
8. Respiration through general body surface by gills in aquatic form tracheae or book lung interstitial forms.
9. Excretion by coelom ducts or Malpighian tubules or green or coxal glands.
10. Sexes are generally separates and sexual dimorphism is often exhibited by several forms.
11. Fertilization is internal
12. Meta metamorphosis takes place in the development.
13. Parental care is also after well market in many arthropods.

MAIN INSECT CHARACTERISTICS

1. The Body is divided into three part, Head, Thorax, and Abdomen.
2. The exoskeleton or the integument is hard and the flexibility is because of the chitin present.
3. The head is formed by the fusion of six embryonic segments.
4. The anterior region of head is provided with mouth parts adapted for bitting and chewing piercing and sucking, siphoning, sponging and chewing and lapping etc.
5. There is a pair of segmented antennae situated near the eyes except the order protura in which they are absent.
6. The thorax consists of 3 segment, Prothorax, and mesothorax and metathorax having one pair of leg in each segment. The second and third thoracic segments carry a pair of wings.
7. The abdomen is comprised of 7 to 11 segments.
8. The alimentary canal is tubular in structure which opens anteriorly into the mouth and ends into the anus.

9. Circulatory system is open type; the blood is circulated in the body through the lateral opening of the heart known as ostia.
10. The Nervous system is composed of the ganglionic masses which are inter connected with nerve cord.
11. Excretion takes places by means of malpighian tubules.
12. Animal unisexually some times viviparous and parthenogenesis.
13. Metamorphosis is common pheromones in insects.

HISTORY OF ENTOMOLOGY IN INDIA

The first general work on insects of Asia was published as "Natural history of Insects" by Donovam in the year 1800. Which was revised by scientists, West Wood in 1842. The natural history phase of Indian entomology received great imptitus with the establishment of following three organizations.

1. The Asiatic Society of Bengal came into existence in 1785. This society encouraged the collection and exhibiting of insects.
2. The Indian Museum of insects was established at Calcutta in 1875 and the Asiatic Society of Bengal was merged with the museum.
3. The Bombay Natural History Society was founded in 1883. A number of its member scientists worked on various insects.

The Govt. of India, therefore, started the publication of the "Fauna of British India" and the first work in this series started with Months of India which appeared in 1889. Maxwell Lefroy was appointed as entomologist to govt. of India in 1903. Lefroy initially worked at Surat for some time and later in 1905, he was attached to the Agricultural Research Institute Pusa (Bihar) as a first Imperial Entomologist. He worked there for about 10 years. There published a very important work on Indian insects in the form of a book named "Indian Insect Life" in 1909. This Monumental book is well known book in the field of Entomology all over the world. A post graduate school was also established there to award (P.G) degree of Master of Science and doctor of Philosophy in Agricultural science including Entomology. The first M.Sc and Ph.D degree from IARI, were awarded in 1961 and the name of Imperial Entomology was redesigned as the Head, Division of Entomology which is still in existance. In the series of entomological work, the Imperial Forest Research Institute was established in Dehradun in the year 1906 by Sir Farther. Earlier, it was started in 1886 as a Forest Rangers Training Centre. The actual construction work of present FRI building started in 1906 and completed in 1914. Dr. S. Nagarkatti was the first coordinator of the project. The Central Institute of Biological Control (C.I.B.C) which was established in 1956 at Bangalore was also taken over by the ICAR on April Ist, 1988 and was brought under the administrative control of National Center of Integrated Pest Management (NCIPM). Dr. H.S. Pruthi, who was the Imperial Entomologists, Govt. of India, became the first plant protection Advisor to this organization. The beginning of the Modern Indian Entomology with the publication of 10th edition of 'Systema Naturae' by 'Carolus Linnaeus' where only 12 Indian insects were included and it forms the first recorded in 1767-1779: "J.G. Koenig" initiated the first regular scientific work on Indian insects and supplied the insect specimens to systematists like Linnaeus, Fabricius, Cramer and Dury. 1782: "Dr. Kerr" published an account of Lac insects 1840: "Rev Hope" published a paper "Entomology of the Himalayas and India"1893: "Dr. Rothney" published the book 'Indian Ants' (Earliest record of biological pest control in India i.e. white ants attack on stationery items kept free by red ants) 1903 : "Maxwell Lefroy" succeeded as the second entomologist.

1906 publication of 'Indian Insect Pests' by **"Maxwell Lefroy**. "T.B Fletcher", the first Govt Entomologist of Madras State wrote a book 'Some South Indian insects' 1914: E.P.Stebbing, the first imperial forest entomologist published "Indian Forest Insects of Economic Importance: Coleoptera. 1937: A laboratory for storage pests was started at Hapur, U.P. 1937: Entomology division was started in IARI, New Delhi. 1939-establishment of locust warning organization after the locust plague during 1926-32.

Contents

1

Body Segmentation of Grasshopper

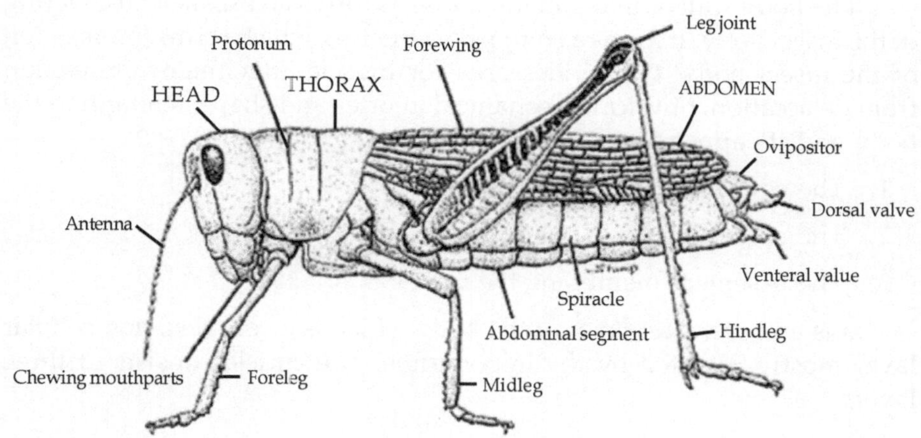

External Structure of Grasshopper

CLASSIFICATION OF GRASSHOPPER

Kingdom	:	Animalia (Animals)
Phylum	:	Arthropoda
Sub-phylum	:	Hexapoda (Hexapods)
Class	:	Insecta (Thorax bears three pair of leg)
Sub-class	:	Pterygota (Wing present)
Order	:	Orthoptera (Hind Wing)
Genus	:	Grasshopper (Hind leg, adapted for jumping, biting and chewing mouth Parts)

Grasshoppers are highly voracious herbivores having the world wide distribution. It is particularly found across – Asia, Africa, Europe and Australia. In India it may be seen in abundance during monsoon season. This pest is of International importance and the following three species are commonly found in India of which desert grasshopper is predominant all over the country.

i) Desert locust (***Schistocerca gregaria*** forsk)

ii) Indian migratory locust (***Lucusta migratoria linn***)

iii) Bombay locust (***Nomadacris succincta***

EXTERNAL STRUCTURE OF GRASSHOPPER

The body wall or integument of insects forms an exoskeletal covering of the insect body. It forms a composite structure which forms the skeleton of the insect body. It provides area for muscle attachment; protection from desiccation, physical/mechanical injuries and shape, strength to the body and its appendages.

1. The cuticle
2. The epidermis
3. The basement membrane the cuticle

It is an outermost layer of the body which is a complex, non cellular layer mostly Secreted by the hypodermis. The cuticle consists of three layers

a) *Epicuticle:* It is an outer most layer, its thickness is 2 μm and consists of protein called cuticle. Epicuticle is build-up, by i) Cement ii) Wax iii) Polyphenol.

b) *The Exocuticle:* It is consisting mainly chitin and protein. It is much thicker than epicuticle. It has phenolic substances which produces hard brown material called sclerotin.

c) *Endocuticle:* It is the inner most layer of the cuticle which contains chitin and proteins; it is thicker than other layer of cuticle.

Function of Cuticle

1. The cuticle provides the proper shape to insect body, and also protects the internal parts of the body.
2. It forms internal organ of the insect body e.g., tentorium apodemes etc.

At the time of development insect casts it cuticle which is known as moulting or ecdysis. The casted cuticle is exuvium is outer covering of the insect body, Epidermis. The hypodermis (Epidermis) are a single epithelium layer of the columnar cells. So many cells are modified into glandular cells; it is also called dermal glands. Theseglands secrete digestive enzymes. Some gland cells secrete seate, spines, bristles etc. which is known as trichogen cells. Trichogen cells are associated with tormogen cells, both are elongated. The hypodermis cells modify into a different type of cells that is called oenocytes. The main functions of these cells are still obscure. It consists of mostly adipose tissue and is the storage site of most body fat.

Function of Hypodermis

1. Hypodermis is a most secreting part of the cuticle which also secretes moulting fluid in the insect body.
2. It absorbs the digestive material of cuticle and also helps to repair the damage cells.
3. Some glands are maintained under which is formed by the epidermis.
   a) Salivary gland b) Silk gland
   c) Lac gland d) Moulting gland
   e) Wax gland f) Mucous gland
   g) Excretory gland h) Poisonous gland.

Basement membrane: Moulting process

Cuticle is hard and forms unstretchable exoskeleton and it must be shed from time to time to permit the insects to increase their size during growth period. Before the old cuticle is shed new one has to be formed underneath it. This process is known as moulting a complex process which involve (Ecdysis 3) Sclerotization :

1. **Apolysis**: [Apo = formation; Lysis = dissolution] The dissolution of old cuticle and formation of new one is known as apolysis. Apolysis starts with repeated mitotic division of epidermal cells resulting in increase in number and size of epidermis, which becomes columnar in shape and remain closely packed. Becauseof this change, the epidermal cells exerts tension on cuticular surface and as a result get separated them from the cuticle. Due to separation of epidermis from the cuticle a sub cuticular space is created and the epidermal cells starts producing their secretion i.e. moulting fluid and cuticular material into this space. The moulting fluid is granular, gelatinous and contains two enzymes viz., proteinase and chitinase which can

dissolve the old cuticle. As the moulting fluid digest the old cuticle, the sub cuticular space increases gradually by the same time and is occupied by the newly formed cuticular layer, the polyphenol layer, wax layer and cement layer into the deposition of definite layers of epicuticle. Procuticle get deposited beneath the epicuticle and subcuticular space is fully occupied. Though moulting fluid is capable of digesting the entire endocuticle, some undigested old exo and epicuticle portions will remain as a layer in the form of an ecdysial membrane.

2. **Ecdysis**: The stage where the insect has both newly formed epi and procuticle and old exo and epicuticle is known as pharate instar. The ecdysial membrane starts splitting along the line of weakness due to muscular activity of the inner developing insect and also because of swallowing of air & water resulting in the distention of the gut. The breaking at the ecdysial membrane is also due to the pumping of blood from abdomen to thorax through muscular activity. After the breakage of old cuticles which is known as exuviae, the new instar comes out bringing its head followed by thorax, abdomen and appendages. Sclerotization: After shedding of old cuticle the new cuticle which is soft, milky white coloured becomes dark and hard through the process known as tanning (or) sclerotization. The process of hardening involves the development of cross links between protein chains which is also known as sclerotization. This tanning involves the differentiation of procuticle in to outer hard exocuticle and inner soft endocuticle. Three types of hormones involved in the process of moulting which are as follows

JH: Juvenile Hormone

Produced from corpora allata of brain that helps the insects to be in immature stage. MH: Moulting hormone: Produced from prothoracic glands of brain that induces the process of moulting Eclosion Hormone: Released from neurosecretory cells in the brain that help in the process of ecdysis or eclosion. Juvenile hormones are a group of acyclic sesquiter penoids that regulate many aspects of insect physiology. JHs regulate development, reproduction, diapause and polyphenisms. JH (Formerly called neotenin) refers to group of hormones which ensure growth of the larve, while preventing metamorphosis.

The Insect Integument

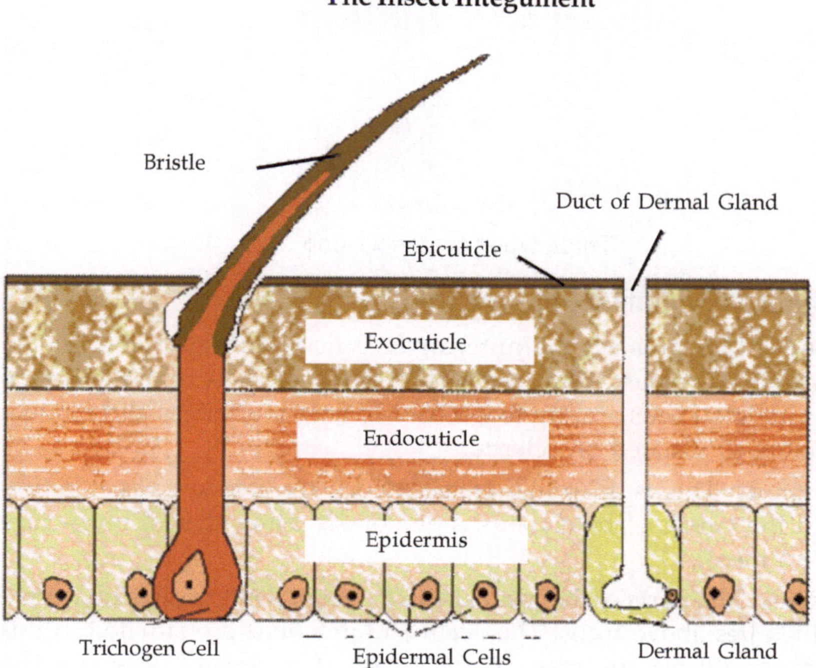

TYPS OF GRASSHOPPER HEAD PARTS

The body of insect is divided into three parts (regions) Head, Thorax and Abdomen.

Head

It is the main part of the body formed by the six segments and attached by the thorax. These are Ocellary anternnal, intercalary, mandibular, maxillary and labial. Head generally assumes the following three definite positions with the body axis.

i) ***Hypognathus:*** In this position the long axis of the head is vertical and mouth parts are ventral. The median line of the head forms right angle with the median line of the body e.g., Grasshopper. Orthoptera (order).

ii) ***Prognathus:*** Here the long axis is horizontal and mouth parts are anterior in position, so the median line of the head is parallel with the median line of the body e.g., Beetle. Coleoptera (order).

iii) ***Opisthognathus:*** In this type, the head is directed backwards and mouth parts are posterioventral e.g., Bugs. (order) Hemiptera.

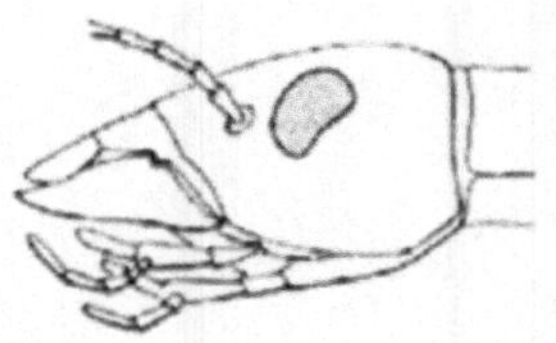
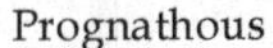

Prognathous

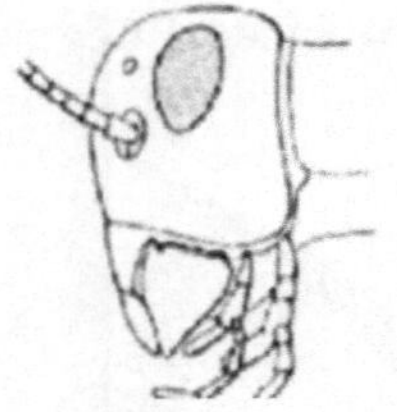

Hypognathous

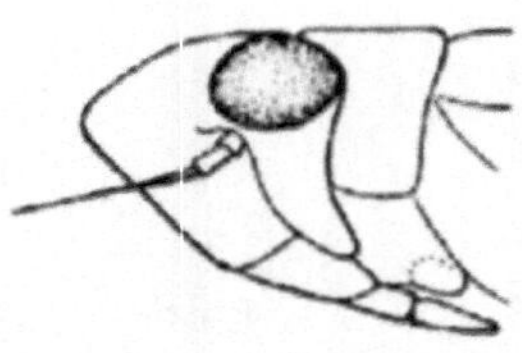

Opisthognathous

Three types of Insect Head

Head Sutures Part of Insect

1. **Frons:** This area on the anterior face lies between or below the epicranial arm just below the vertex, which extends from the frontal suture to clypeus and up to the base of both mandibles and median ocellus, is situated on sclerite.
2. **Clypeus:** This is a lip like area between the fronto clypeal suture and the labrum. The clypeus is one of the plates that from the face of an insect.
3. **Vertex:** The parts of the head above the frontal suture are the vertex, which lies above frons. The ocelli and antennae are situated in this region. Vertex an area on the head of an insect between the compound eyes the top of the head.
4. **Epicranial suture:** It is an inverted Y shaped suture, lies posterior between the eyes, whose stem begins on the back of the head crosses the vertex and fork on the face.
5. **Genae:** These are the lateral walls of the Head situated below and behind the eyes.
6. **Compound eyes:** It is paired structure situated in dorso-lateral position to head capsule. Each eye is surrounded by a narrow ring like ocular sclerite.
7. **Ocelli:** These are located in between the compound eyes and generally three in numbers.
8. **Antennae**: These are situated in antennal socket between the compound eyes.
9. **Tentorium:** The endoskeleton of head is known as tentorium. Thus tentorium is composed of four principal parts namely, the anterior arms, posterior arms, corporatentorium or central mass and dorsal arms. The tentorium (Pleural tentoria) is a term used to refer to the frame work of internal supports within an arthropod head. The tentorium is formed by in growths of the exoskeleton, called apophyses, which fuse in various ways to provide rigid support for the muscles of the head.

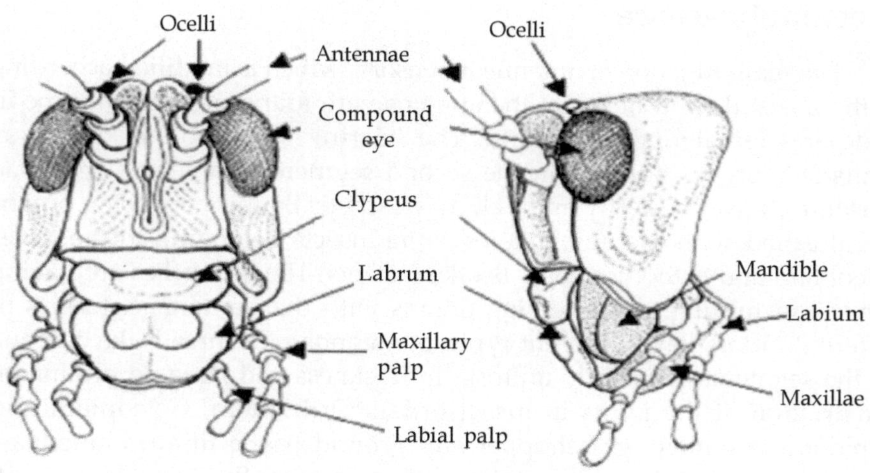

Head Sutures Part of Insects

Head Appendages

The Head Appendage includes the antennae and mouth parts.

Antennae

These are multi segmented and may be divided into three Parts.

1. **Scape.** This is basal segment of antennaly, and attached with head.
2. **Pedicle.** It is the second segment of antenna which is shorter than scape it bears sensory apparatus known as Johnston organs.
3. **Flagellum.** This part is also known as clavola.

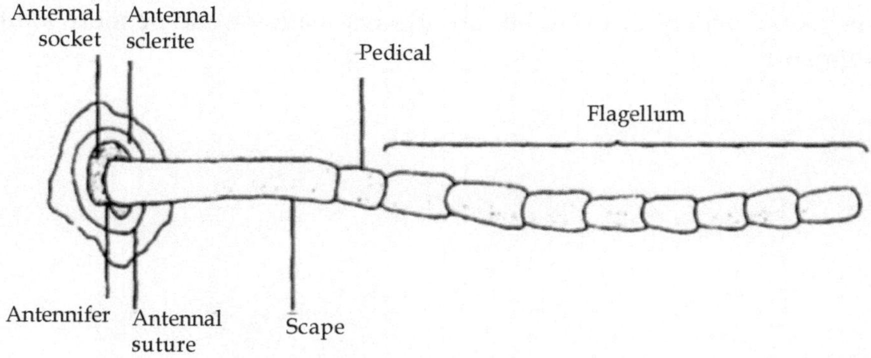

Typical type of antennae

Function of antennae

The main function of antenne is sensory, which is modified according to the use and need of insect the antennae are found of different type in male and female e.g., mosquito. The hearing organs often known as Johnston's organs situated in the second segment of antenna e.g., male mosquito, green butterfly and yellow wasp etc. Sound producing organs are attached with the antennae of some insects belonging to the order coleoptera and orthoptera e.g., Beetle, crickets. The butterflies are having some transmitting and receiving organs with their antennae. Insects of different orders have different types of antennae. In thread like antenna all the segments are nearly uniform in thickness and have no prominent constriction at the joints in insect orders Orthoptera, Coleoptera and Lepidoptera namely grasshopper and ground beetle filiform antennae. Setaceous one point tapering antennae e.g., dragon fly. *Moniliform* – All the segments of this antenna are globular in shape and of uniform thickness looking like a string of beads e.g., Isoptera order and insect termites. Pectinate antennae (comblike structure sawflies). Clavate the distal segments gradually increase in diameter e.g., butterflies. *Capitate* .The distal segments suddenly increases in diameter e.g., Khapra beetle. *Geniculate* - In this type of antenna, the first segment is long, second is short and flagellum is made of small segment which are bent on the scape just like a bent knee order Hymenoptera-Honey bees. *Fusiform*-the basal and distal segment of this antenna are smaller and thin while middle segments are larger just like a radish. The last segment is modified into hook like structure order Lepidoptera-insect mouth etc.

Antennae (Singular: antennae), sometimes referred to as "Feelers" are Paired appendages used for sensing in in arthropods. Antennae are connected to the first one or two segments of the arthropod head. They vary widely in form, but are always made of one or more jointed segments.

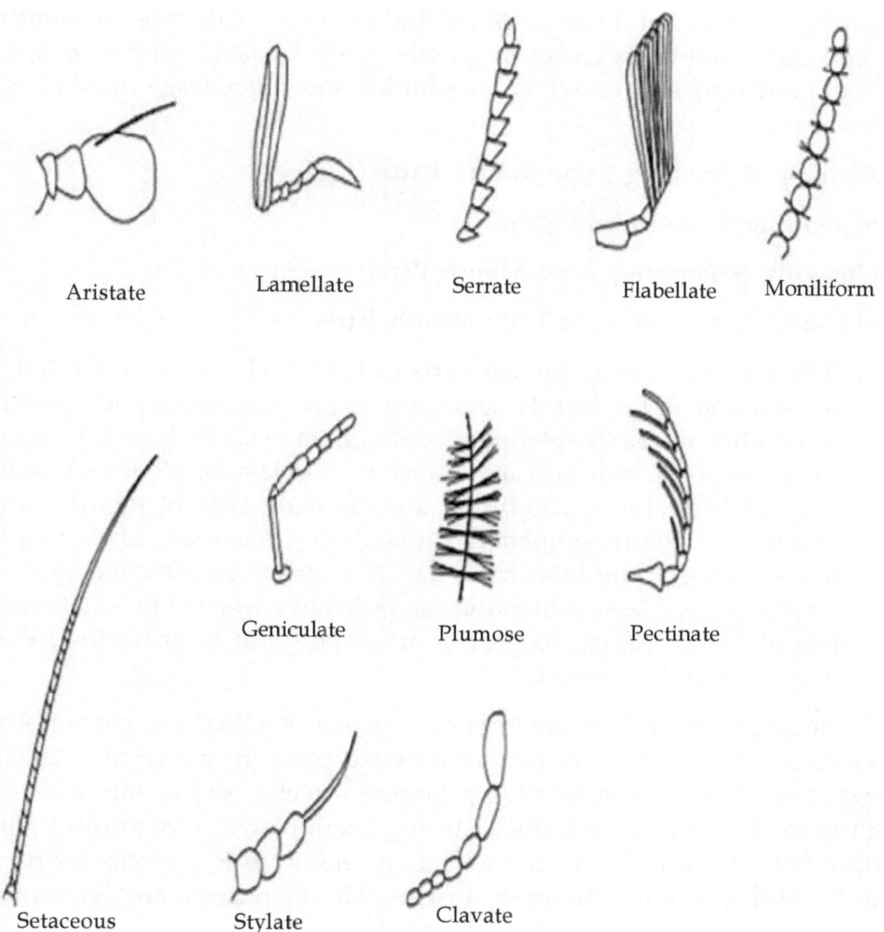

Modification of antennae of insects

Mouth Parts of Insects

The insect mouth parts are considered to be the diversed forms. Fundamentally it (mouth parts) includes Hypopharynx, one pair of mandibles, one part of maxillae, posterior lower lip-labium. The mandible, maxillae and labium represent modification of typical paired appendage of three pairs of body segment. The typical anterior part of insect mouth is labrum. The mouth parts of insects can be divided into different groups depending upon the type of the food and feeding. The insect mouth part is given below:

A. **Biting & Chewing Type or Mandibulate Type:** This type of mouth parts are found in cockroaches, grasshoppers, locusts, termites, wasps, book and bird lice, earwigs, dragonflies and other large number of insects.

B. **Piercing & Sucking Type Mouth Parts**

C. **Siphoning Type Mouth Parts**

D. **Chewing & Lapping Type Mouth Parts**

E. **Haustellate or Sponging Type Mouth Parts**

1. **Biting and Chewing Mouth Parts of Insect**: These types of mouth parts found in the insects belonging to phylum Arthropoda under order **Orthoptera, Isoptera** and **Coleoptera** etc. The insects having such type of mouth part make hole on the leaves, seed, and fruit and cut irregularly. The biting and chewing type of mouth part consists of labium, epipharynx, mandibles, maxillae, labrum and hypopharynx. The labrum is flab like structure attached to the clypeus upper lips. It functions as protective upper lips and inner lips of mouth cavity, to protect mandibles and to grind the food into the mouth of insect.

The Epipharynx: They are important structure with the upper surface of labrum. The epipharynx part bears taste buds in mouth of insects. Mandibles - A pair of mandibles is located directly behind the labrum. Each mandible segmented, thick strong, triangular in structure upper surface flat. The mandibles are movable in mouth. It is a protractor two muscles abductor and adductor muscles. The mandibles are primarily meant for chewing and grinding the food.

The maxillae: The maxillae are used for holding the food so that mandibles may perform their functions easily. Maxillae are a paired structure lying behind the mandibles. Maxillae are divided in different part as discussed below:

i) *Cardo:* It is a triangular basal part attached to the head capsule and regulates movement of maxillae.

ii) *Stipes:* It is a rectangular structure in maxillae situated above the cardo. There is a distinct sclerite known as palpifer to which the palpus is attached.

iii) *Galea:* They are outer lobe soft and segmented. The basal segment is termed as a parastipes.

iv) *Lacinia:* It is a inner lobe of maxillae. The inner side hard spine or teeth mallize in maxillae surface.

v) *Maxillary Palps:* It is a antennae like appendage attached with the lateral side of the stipes. The maxillary palp is attached as palpifer. Maxillary palp is a five segmented and sensory in function.

Labium: It is a lower lip of insect mouth cavity the labium appears to be a single organ, but consists of two parts which have fused to form a functional structure. It is divided in different parts as i) Submentum, the flat leaf like basal part of the labium. ii) Mentum central part of the labium which bears a pair of three segmented pulp on its either side. iii) Prementum upper portion of labium, which has four terminal lobes. The median pair is glossa and lateral paraglossa. These lobes are attached to lacinia and galea of the maxillae and are commonly known as ligula.

Hypopharynex: It is a tongue like structure in the mouth cavity. The salivary gland is believed to open in the mouth cavity of insects via hypopharynx.

2. **Piercing and sucking type mouth parts of insect:** These types of mouth parts are found in mosquitoes and bugs. These are highly specialized mouth parts which are adapted for piercing and sucking sap from animals and plants epidermis. Bug type is examplinary type of mouth part found in order Heteroptera and Family Pentatomidae which are used for sucking sap from the plants. It is a different part of mouth, paired mandibles, maxillae, labrum, epipharynx and Hypopharynx all six parts are modified to needle like style and broad tubular like labium. The labrum-epipharynx, formed by the fusion of labrum and epipharynx, is the dorsal most stylet covering the opening of the groove of proboscis. Hypopharynx is somewhat flat and double edged sword stylet covers the food channel and has a salivary duct inside. Needle like mandibles lie on each side of the labium, epipharynx, the distal and being used for sucking sap from human skin and other animal's skin. The maxillae are also needle like, distally serrated and located laterally hypopharynx in the groove of the proboscis. When the piercing is over, the dilator muscles contract and upward suction is created through the food channel by capillary action as a result of that the sap is sucked up from the plant cells by the turgor pressure into sucking chamber.

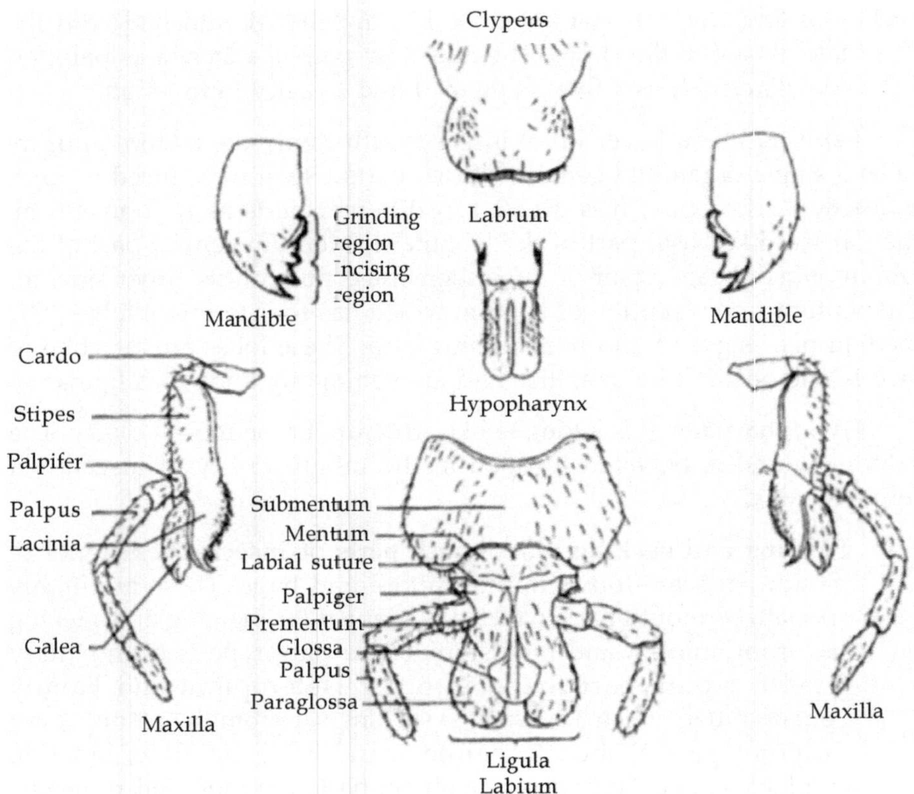

Chewing and biting type mouth parts of Insects

Mouth Part of Sponging Type

The sponging type mouth parts are commonly found in the order Diptera. It is found in non-blood sucking insects. The example is house fly. These mouth parts are highly modified and primarily used for feeding of liquid food soluble in saliva. It is long labrum tapering and flap which originates from the distal margin of the rostrum and covers the anterior surface of the haustellum. The mouth is situated medially at the base of the labrum and hypopharynx. Mandibles are absent. Maxillae palp is single segmented and labium is fully developed and modified into long proboscis. It has three parts. (A) Rostrum. They are two small lateral sclerites which support a pair of single segmented maxillary palp of insect. (B) Haustellum is cylindrical shape and demarcated by a distinct long medio dorsal labial groove which work as the food channel. (C) Labellum distal portion of the labium and is also known as oral disc.

Chewing and lapping mouth parts: This type of mouth part are found in the class insects of order Hymenoptera e.g., Ants Honey bees and Wasps. These are generally used for collecting the pollen and nectar from the flowers. The mouth part under labrum and mandibles are just like chewing type of mouth parts. Mandibles on the lateral side of labrum spoon shaped mainly adapted for molding the wax and for comb formation. Maxillae are either side of the labrum. They are maxillae and labrum form flat and elongated structure and long blade like stipes which the glossa form an extensible channeled organ with a small labellum at the tip. This proboscis is used to suck into nactaries of flowers.

The Siphoning type mouth parts of insect: The siphoning type mouth is found in order lepidoptera under insect butterflies and all moths. The proboscis is coiled type of moth. They are mainly used for sucking nectar from flowers. The labrum is not fully developed and narrow transverse band at the lower edge of the clypeal region of the face labrum. The labrum on its lateral sides has small hairy lobes. The mandibles are absent. The mandibles are usually reduced. Its labium is undeveloped. This is represented only by segmented labial palp and a very small basal plate.

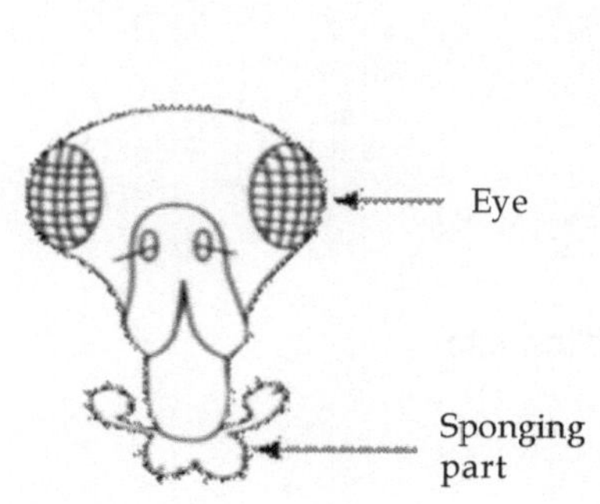

Sponging type mouth parts

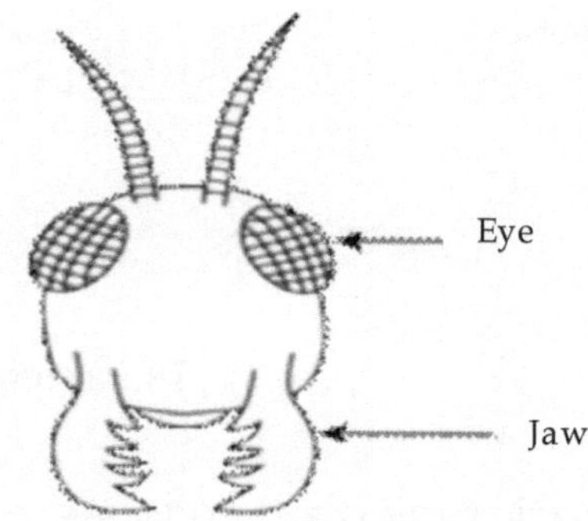

Chewing (beetle) type mouth parts

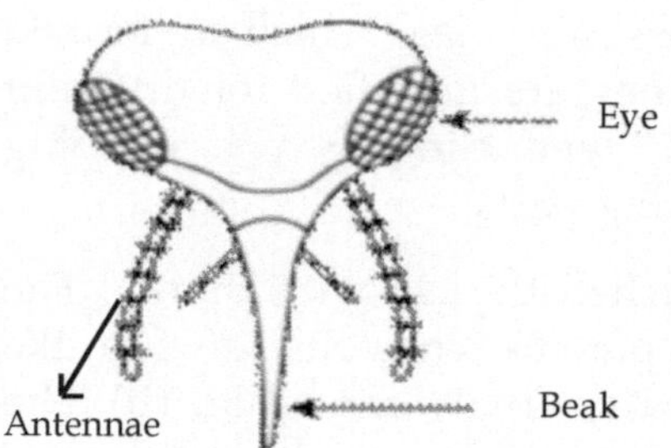

Piercing sucking mouth parts (mosquito)

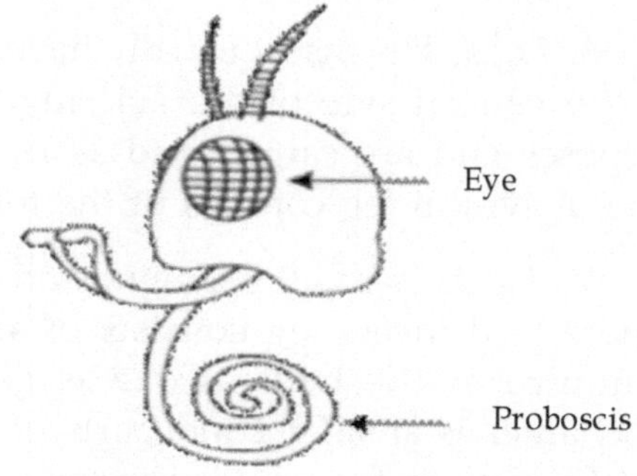

Siphoning mouth parts

Thorax regions of Insects

The insect thorax region situated between.

Head and abdomen: It is divided into prothorax, mesothorax, and metathorax. In some insects like those of orders orthoptera, Dermaptera and odonata there is a very small region between the thorax and head known as cervix region of insect. Some time it has also been referred as. the neck region of the insect. Each thorax segment bears a pair of legs. The first pair of wing is attached in the mesothorax, where as the second pair is found in the metathorax. In winged insect the meso and meta thorax are so closely attached to form a prothorax. The thorax is armed with various sclerites which are termed by their specific location as tergun pleuron located onThe insect thorax region situated between head and abdomen.

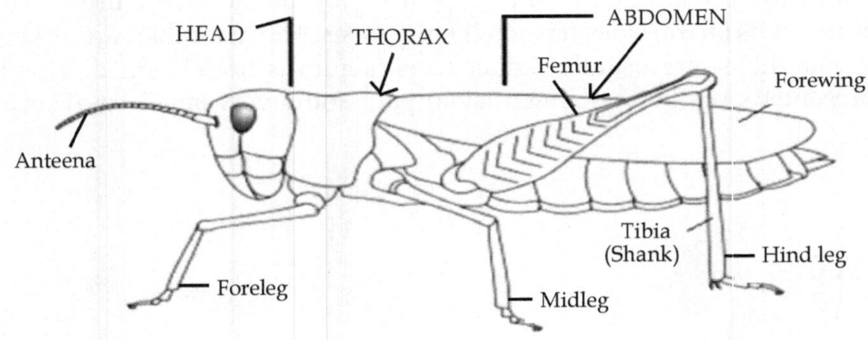

Thorax regions of Insects

Thoracic region appendages

It is a two type: **(A)** Legs **(B)** Wings

A. Legs: The insect usually have three pair of legs which are located on the ventral side of thoracic region they are modified for different purposes and are catagorised as running type, clinging type, jumping type. A typical leg consists of the following parts.

(a) Coxa is basal segment and attached the leg with thorax. The generalized insect leg consists of six segments, coxa fits the cup like structure of the body. Coxa is generally freely movable. (b) The Trochanter is small second parts of the leg which articulate with coxa and fixed with femur. (c) Femur is a largest and powerful part of the insect leg. (d) Tibia is slender like structure, usually long and provided with downward projecting spines into which helps in climbing. (e) Tarsus is a fifth segment of the insect leg. It is divided into two or five segments. (f) Pretarsus is the last terminal segment of the leg which is represent by complex set of claws which bear one or more pads. They have undergone many modification and have been adapted to a wide variety of function including swimming, prey capture, pollen collection and digging.

a) *Cursorial Legs of Insect*: These are almost similar to that of walking legs and are adapted in such a way to avoid slipping, e.g., Ants, earwigs cockroach.

b) *Ambulatorial Legs of Insect:* It is the generalized from of an insect leg which is usually adapted for walking, e.g., grasshoppers.

c) *Clinging Lege of Insect:* The clinging legs are smaller and flat. The clinging legs are found in lice etc. The Tarsi is single segmented and each ends in a powerful claw which works against a tibial process e.g., louse.

d) *Pollen Collecting Legs of Insect:* The hind legs of the honeybees are adapted for carrying pollen from the flowers. The coxa, trochanter and femur are normal inshape, while the tibia of hind legs is dialated and covered with long dense hair thus forming a pollen basket which is primarily for storing the pollen grains parts of flowers. The first tarsi segment is flat and its inner surface is densely closed with several rows of short stiff spines forming a brush known as scopa. It helps the bees in collecting the pollen adhering to the hair of its body.

e) *Jumping Legs of Insect:* This type (saltatoriat) of legs is developed in grasshopper and beetle where the femur of the hind leg gets enlarged and accomodates the powerful tibial muscles. These muscles help in jumping the insect trochanter muscles.

f) *Sound Producing Legs of Insect:* These sounds are made in two genendciolyer tances legs typically adapted for producing sound where in the femur of hind leg of male insect like grasshopper or cricket is provided with the row of file on its inner side. These femoral file work against the outer surface of each tergum or costal margin of the forewing thereby producing a sound. Many beetles produces sound usually by rubbing one pars of the (a scraper) against anther part of the file.

g) *Natatorial Legs of Insect:* The legs of several aquatic insect are modified in such a way that they facilitate swimming e.g., water beetle. The coxa and trochanter are simple and smaller whereas the tarsus is five segmented and last terminal segment is triangular and pointed e.g. Jaint water bug.

h) *Suctorial and Sucking Type Legs of Insect:* The Tibia is comparatively smaller and flat.

Tarsus is five segmented and first and third segment modified into a swollen ball like structure and provided with stiff spines at their margins. They are also helpful in sticking with grass and leaves. These legs are also developed for sexual purpose in which the coxa and trochanter of the first legs are small.

INSECT LEGS

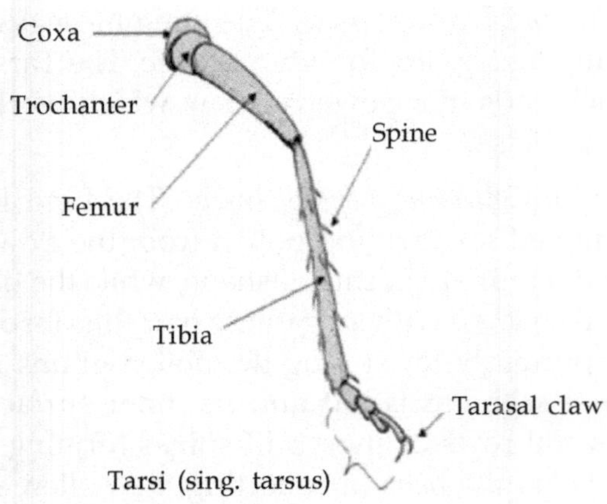

Typical type leg of insect

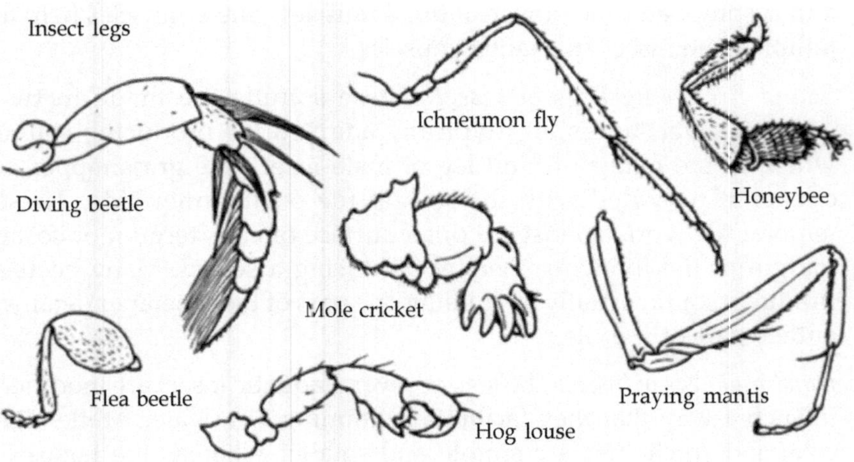

B. Wings: Generally two pair of wings is present which are the out growths of the body wall along the lateral margin of thorax. The presence of wings is one of the most characteristic features of the insect and provides one of the most useful Taxonomic aids of insects. The wings are of course the organs of aerial locomotion in most cases, but, like the legs they have undergone extensive adaptive modification in Insects. In order diptera there is only one pair of wing and the hind wing is modified into halteres.

The Elytra: In the insects of order coleoptera the forewing gets much hardened to form horny sheeth which protect the outer membrane and inner membrane of hind wing of insect. Hemely tra this is well illustrated in hemiptera insect where in the fore wings are thickened at their basis like elytra and remaining as soft membraneous at distal parts that is why they are frequently termed as hemelytra. Halteres–the hind wings of order diptera (housefly) are modified into knobbed Thread like balancing organs known as halter. Halter is a balance organ of House- fly. Tegmina is found in the fore wings of order orthoptera and Phasmida where they are hardend and Leathery in consistency.

Wing structure in insects: The wing appear as the thin or transparent darkly fan like flattened membraneous structure. The basal sclerites help in articulation of wing with the thorax as they rotate with one another. The wing bears a specific pattern of venation which is derivedfrom unique arrangement of veins. The veins provide mechanical support and folding to the wing. Theys play important role in determination of direction of the wind. The Wing is movement and navigation during flight.

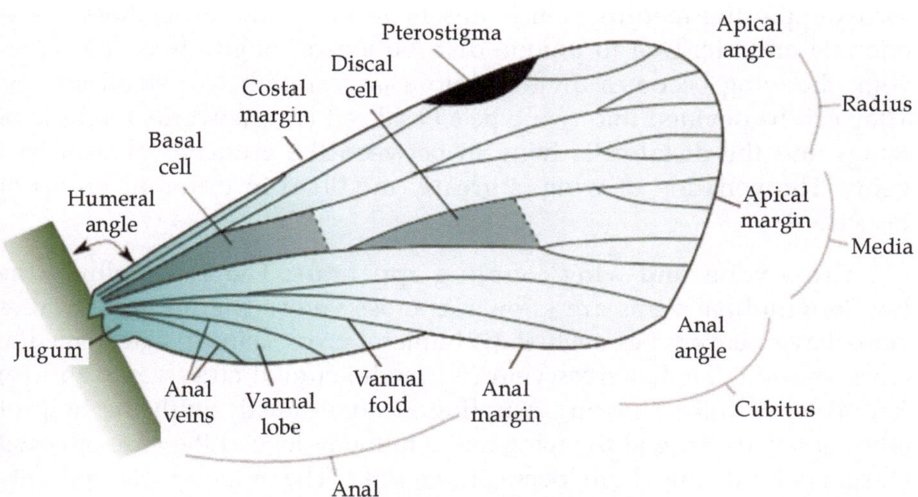

Structure of Wing venation in insects

The wing regions and basement of the insect: The margins of wings are named as the anterior margin and posterior margin. In order diptera a pair of membranous lobes at posterior margin of the wing base is predominantly evident and is know as the outer and inner squamma. It is well developed in the housefly. They include the tegula, the axillary cord, axillary sclerites and variable form of skeletal plate. The tegula are mostly confined to the forewings, particularly in the lepidoptera, and diptera order each lies at the base of the costal vein as a small hairy chitinous pad of dorsal side of insect. There are three axillary at the base of wing. The second axillary lies at the base of radius vein and articulate with the first axillary proximally and with the wing process of the pleuron end. Third and fourth articulatate directly, connect with the posterior notal wing process. It is in present axillary orthoptera order. The wing venation of Insect Bugs Pentatomidae. Generally in all the insects there is some similarity in wings venation and therefore, it is presumed that all types of wing venation have developed from the common base or the same ancestor. By means of an extensive study of wing venation in different group of insect. Accordingly we say Primitive wing venation has developed from two tracheae which are situated on the anterior and the posterior basal margin of the wing and their branches are spread all over the wing. Each main trachea gives rise to three principal veins, thereby forming six principal veins namely costa, radius, medius, and cubitus, penultimate and ultimate. Each principal vein gives rise to a sub-vein near its base. The principal veins are denoted by + sign and sub vein represensed by - sign. Thus the whole wing venation system is represented by + and - sign. Such type of hypothetical wing venation in never met in any insect as one or the other vein is invariable found lacking for example the medius vein is absent in some like order hemiptera, odonata and etc. Due to unique distribution of longitudinal and cross vein, the wing becomes divided into a large number of small spaces. They can be devided into two type's basal cell lying towards the base of wings and the distal cells lying in between the branches of principal veins. The venation of wing is greatly modified in different group of insects.

Cross veins and wing coupling apparatus: The veins joining the two longitudinal veins are known as cross veins. The important cross veins have many types such as (i) humeral cross vein (ii) Radio-medial cross vein (iii) Medial cross vein (iv) medio-cupital cross veins and (v) Radial cross vein. Thewing coupling is done mostly with the help of lobes or spines lying at the wing and a humeral lobe at the base of costal margin of hind wing. Both lobes contain setae. The humeral lobe specially bears the frenular bristles. From this primitive type of wing coupling mechanism, complex types have been evolved. Other insects have the wings coupled by more distal modification which hold the costal margin of the hindwing to the anal margin of the forewing. Thus Hymenoptera, Heteroptera have a row of hooks, the hamuli, along the costal margin of the hind wing catch into a fold of the forewing. Thus synchronus action of the fore and hind wings, there is enabling the insects to fly more swiftly. Types of various coupling apparatus are: Jugal and Humeral lobe, frenulum and Retinaculum and Hamuli.

Abdomen of Insect: The abdomen is generally comprised of 10 rings like segements which are also known as uromeres. They are almost similar in structure except of those on the posterior end which are highly modified to form the external genitalia. The Abdomen of grasshopper consists of 10 segments the dorsal surface is composed of a thick cuticular layer known as tergum which form the 2/3 portion of each of the segment and termed as sternum. In male and female

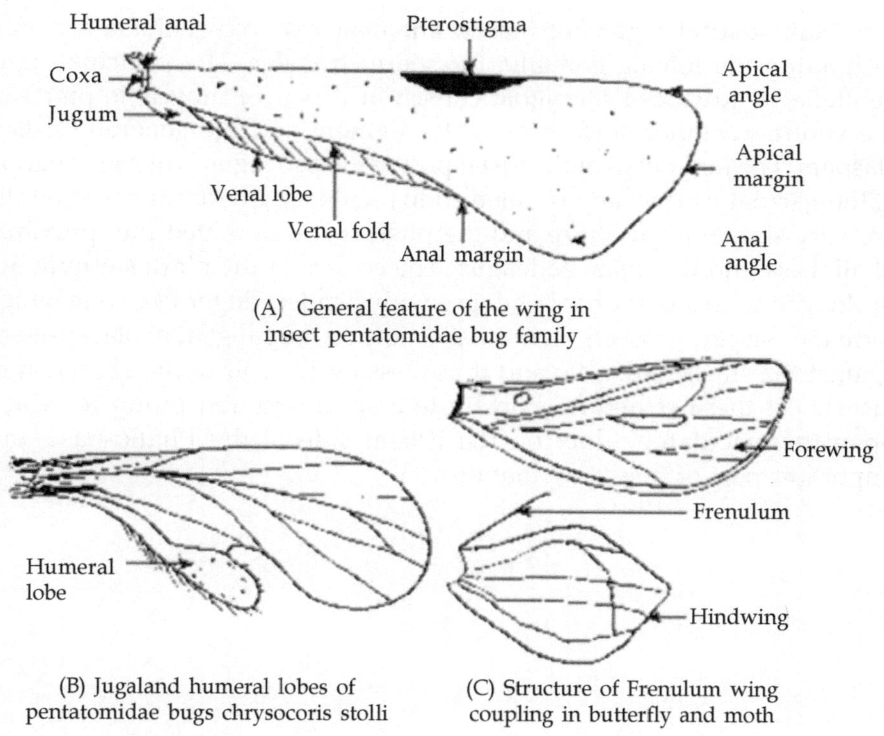

(A) General feature of the wing in insect pentatomidae bug family

(B) Jugaland humeral lobes of pentatomidae bugs chrysocoris stolli

(C) Structure of Frenulum wing coupling in butterfly and moth

Grasshopper 9th and 10th abdomen tergites are so closely united that they appear to be one segment. In the generalized female pterygote insect family pentatonic, modified him eight and ninth abdominal segment appendages of form the ovipositor or egg-laying apparatus, which is composed of two pair of basal valvifers. The female gonopore is usually on or posterior to the eight The male external copulatory organs penis-aedeagus is usually borne on the ninth abdominal segment .Female Genital Organs of Insect. The genital of female grasshopper consists of a copulatory pouch and a special egg laying organ known as ovipositor. These organs are considered to be the appendage of 8th and 9th abdominal segments. The ovipositor helps to the female in depositing her eggs in the soil by digging hole in the ground. The inner valves are considered to be vestigial and non-functional where the dorsal and ventral valve fit together to from the functional ovipositor. There are three pairs of valve which collectively form the ovipositor. The first pair originate from the 9th sternum which is triangular pointed and hard in structure known as dorsal valvies. The second pair of ventral valves originating from the 8th sternum, which is similar to dorsal valve and projected updown and a 3rd pair of valves lying in between the dorsal and ventral valves known as inner valves.

Male genital organs of insect: The male external genitalia are used in holding the female genitalia for sperm transfer. The principal male genitalia in advanced pterygote consist of a pair of moveable plates at the ventro-posterior surface of ninth segment and are generally called claspers. The inner wall of the distal part of the aedeagus is in continuation of the ejaculatory duct and is called endo phallus.In thysanura the genitalia are very simple in structure and the phallus is differented into proximal phallobase and the distal aedeagus. The coxites of the ninth sternum are prolonged in to a pair of appendages with slender, finger like style which form the clasper of insects. In normal condition the subgenital plate presses against the supra and plate and thus clossing the end of the abdomen of insect. All these structures constitute a spern ejection pump in which open the ejaculatory duct on the dorsal side of the Phallo-base and important part of insect copulations.

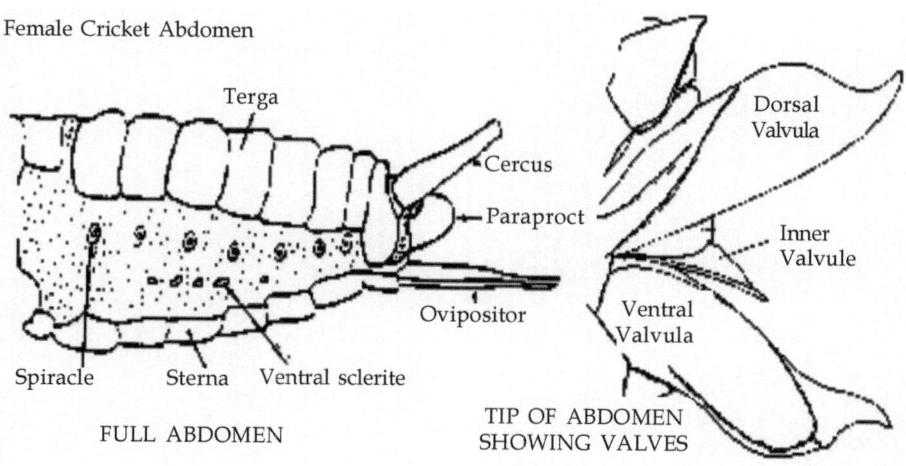

(A) Female genitalia of grasshoppers

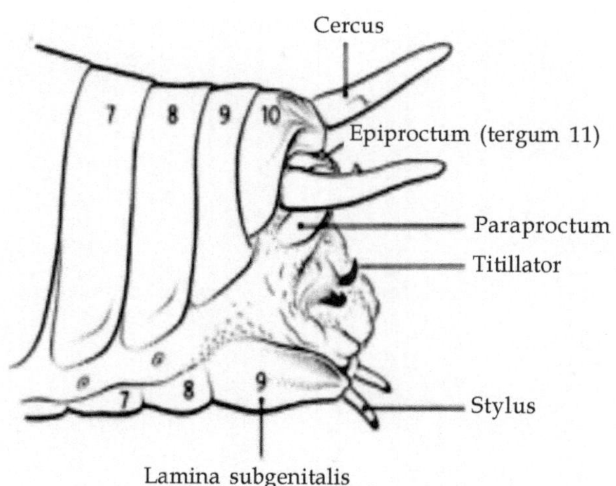

(B) Male genitalia of grasshoppers

2

The Division of Insect Integument

The integument is composed of the cuticle and under lying epidermal cells that secrete the cuticle. The integument is a complex vulnerable organ system of diverse function. The insect outer body covering is called integument. The integument is a central subject intimately related to a variety of applied - Problems such as the mode of action, insecticides, ecology, cuticle and endocrine glands present in the body of insects. The insect integument consists of 3 parts and plays crucial role in locomotion, breathing and respiration, feeding, excretion, protection from desication behaviour of insect.

1. Epidermis Layers
2. Basement Layers
3. Cuticle Membrane

1. **Epidermis Layers:** An insects exoskeleton serves not only as a protective covering over the body, but also as a surface for usual attachment, a water tight barrier against environment and much more. Epidermis forms the insect outer layers of body covering, usually thick and ectodermal in origin. The plasma membrane of the adjacentment cells is joined by septate desmosomes to form a functional syncytial layer. The center of the diffusion of molecular transfer is via cytoplasmic plasma membrane. The cells contain numerous mitochondria, golgi vesicle, endoplasmic reticulum as well as cytoskeletal membrane. It may secrete the basement membrane also, contains other types of dermal glands. *E.g., Exocrine glands, moulting fluid* etc. During moulting it secretes moulting fluid which dissolves the old (endocrine gland) endocuticle before the immature insect moults. It is a multi-layered structure with four functional region epicuticle epidermis and basement membrane.
2. **Basement Layers:** The basement membrane is a supporting bilayer polysaccharide and collagen fibre. The outermost non-chitinous layer

of the cuticle. The basement membrane contains neural cells secreted by haemocytes cells.

3. **Cuticle Membrane:** The cuticle is made up of two layers: The epicuticle which is a thin waxy water layers and Endocuticle which is comparatively thicker and untanned.

The insect cuticle is a complex, outermost layer secreted by epidermal cells and is the seat of several metabolic activities. They are outermost non-chitinous layer of the cuticle. The epicuticle is exceedingly complex and extremely important layer of cuticle. It is 4 layered made up of wax canal that contains wax filament. The wax canals are transporting wax molecules to epicuticle via pore canal from their secretion. An outer cement layer of unknown composition is secreted by dermal gland. That determines the surface properties of the cuticle will be water repellent or water attractant. A waxy bloom appears on the surface of cement layer in some insects. The wax layer consists of many ordered monolayers of lipid directly associated with the cuticulin layer and is responsible for growth of many cuticle layers. It is quite impermeable to water and is very resistant to acids and organic solvents and it covers the entire integumental surface including the tracheoles and duct of insect scant glands. The cuticle layer is about 4 mm thick and appears as homogeneous layer. Outer cuticle is a divided in two layers of exocuticle and endocuticle. The exocuticle is thin and soft layer of insect body. It is a pigmented layer but endocuticle is composed of successive light and dark layer which correspond to daily growth layers. The dark and light layers more lamellar pattern and are considered to be the result of the orientation of layers of microfibrils of chitin and probably protein embeded in a protein matrix. The pore canals are tiny tubes twisted vertically and extend from the epidermal layer nearly to the external surface of the epicuticle.

The chemical composition of the cuticle includes polysaccharide chitin and various structural proteins and other cuticular constituents. Chitin is a higher molecular weight polymer of anhydro-N-acetyle-D glucosamine and D-glucosamine linked through 1, 4, B-glucoside-bonds mostly in pro-portion of 91. It may make upto 25 to 60% of the dry weight of exo- and endo cuticle. Chitin is absent in epicuticle. It does not exist in a pure state naturally, but is combined with a protein as a glycoprotein.

The Insect Colouration

The insect colouration is commonly produced by the complex mixture of different pigments and the same colour may be achieved indifferent ways. The very common pigments are melanins (yellow, brown, black), carotenoids (red and yellow) development from plant tissue, and ommochromes (red) found as eye pigments, chlorophyll derivaties (green) haemoglobin (reddish) flavins (greenish yellow). The insect colour may be due to the various pigments present in the cuticle, scales, epidermal cells their combinations etc. Method of Insect Moulting.

Moulting is the process by which insects grow, generally accomplished through the early years of the insect's existance, molting allows the body of the insect to expand under controlled, protected conditions. During moulting epidermal cells separate from the old cuticle and begin to secrete the new. Then the epidermal cells secrete moulting fluid that contains chitinase and protease digesting about 80-90% of the endocuticle. The insect moulting apolysis is a thin homogeneous, transparent excuvial membrane that appears between epidermis and old cuticle. It is resistant to moulting fluid. The exo and epicuticle are also resistant to the action of the moulting fluid and make up the portion of the integument that is shed at ecdysis. This process is facilitated by ecdysial line beneath which only epicuticle and endocuticle are present. Since the endocuticle is digested during the moulting process a line of weakness develops. This process is repeated several times during the life span of the insect depending on the species.

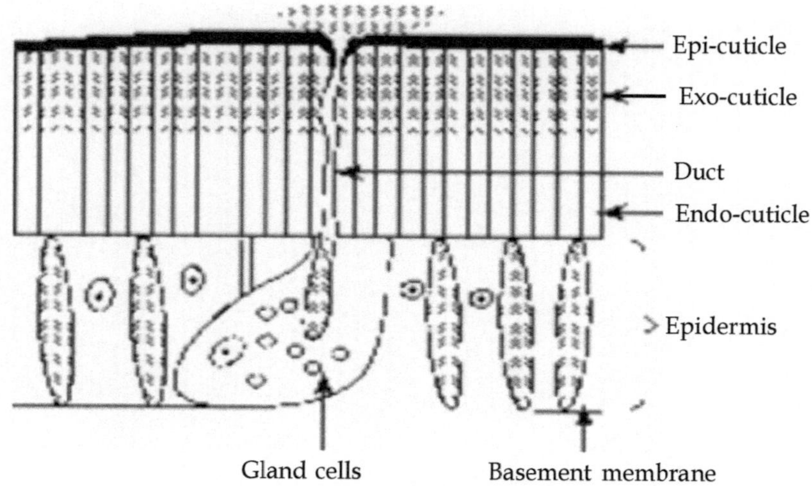

Structure of integument of insect

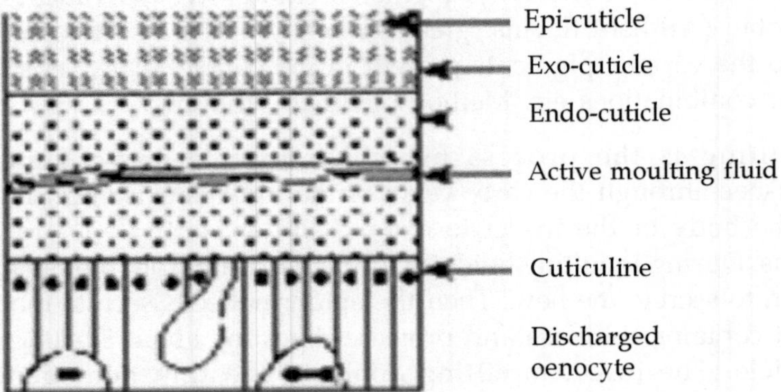

Insects of moulting methods insect

3

Insect Origin and Evolution

ORIGIN OF INSECT

Entomology is a branch of zoology which deals with insects. In this branch we study the insect origin and evolution of insect and their diversity and classification and other parts of insect body. Insects belong to the class insecta in the phylum arthropoda, the largest group in the animal kingdom. All arthropods are characterised by having segmented body, claws, chitinous exoskeleton, ventral nervous system and dorsal heart. The class insecta is generally considered to have evolved from myriapod of some sort during the Devonian period. Based on different mandibles and mandibular movement concluded that insects are not direct insect group myriapods the mouth is located between the prostomium and the first body segment and the anus open in the periproct.

The first great step was the development of a pair of ventral appendage. Insect legs on each body segments and locomotion, parallel to this an improvement in the sense organs of the head occured eyes and Antenna were the ultimate result this level of organisation is met in onychophora. A series of diagramms is helpful in a very simplified way in visualising the hypothtical origin of insects.

Next step represented by figure is the development of articulation in the legs improving the locomotion. The first pair of legs is considered to be used in pushing food into the mouth at certain geological period during evolution. Eyes and antennae are well developed can be looked upon as representing the myriapod level of body organisation, bi-lateral appendages of segment 3, 4, 5, 6, 7, becoming the typical primitive mandibulate mouth parts. Appendages of second body segment ultimately become the antennae, those of the fourth become the mandible in insect and fourth fifth and sixth segments become the first maxillae and second pair of maxillae (labium). These three segments are termed as genital segment of last insect. The Appendage of 5th, 6th segment became the first and 2nd pairs of maxiallae. The genatial segment became consolidated with the prostomium resulting in a head structure of compound origin much like those insects and their allies. The body has been differentiated into the three tagmata head, thorax and abdomen of insects. The appendage of segment 7, 8 and 9 have been retained as locomotor structure, the abdominal segment 8 and 9 has been modified as external genitalia, cerci and the appendages of segment 11th have been retained

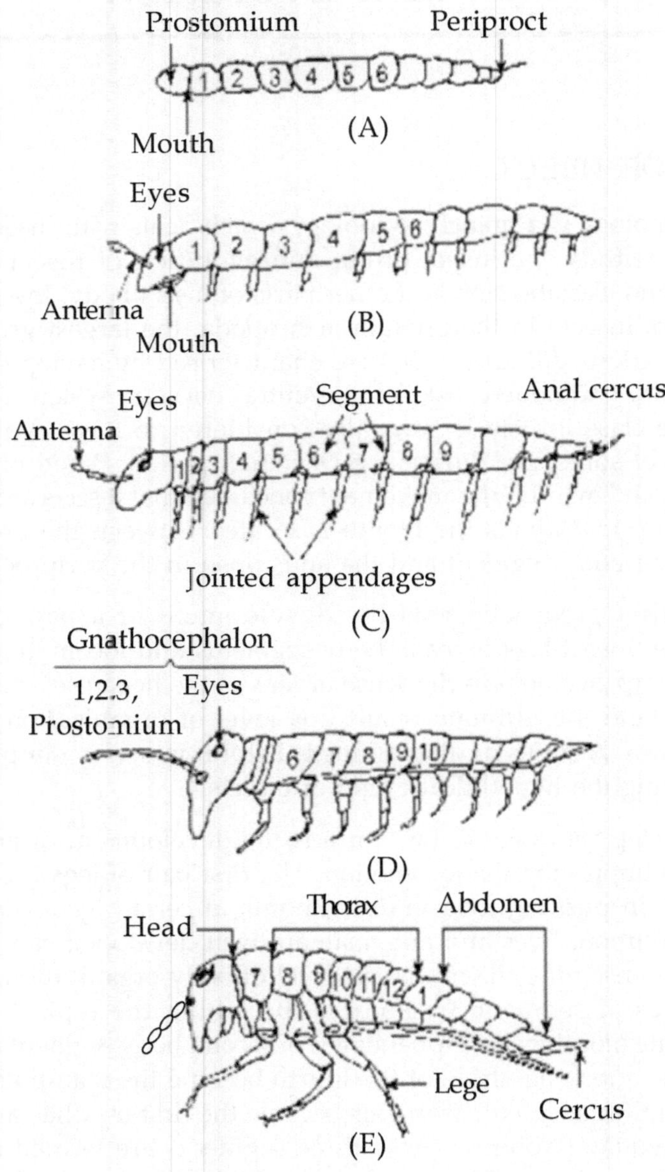

Hypothetical stage in the evolution of insect

Evolution of Insect: The evolution insect dates back to the Devonian period, with the oldest insects.

1. **Insect Evolution and Wingless Insect:** The first was the appearance of primitive wingless insect which probably resemble contemporary order-thysanura under insect silver fish order - Thysanura. These primitive apterygota are thought to have arisen during the Devonian period. Thus they are grouped in separate subclass from the winged insects called apterygota. Insect can be classified into groups depending on whether their mouth parts are exposed. Other Diplura evolved from ancestor with the same body plan on that of winged insects the group Ectognatha. The other two oder of wingless insects, collenbola and pratura, differ from this body plan and manyconsider them to be sister groups to insecta rather than true insects.

2. **Wings Origin of Insect:** The concentration of walking appendages in the midbody region the thorax was undoubtedly an important pre-requisite for the later development of wings, for this required the evolution of a boxlike well musculated thorax. The insects having wings and those that have lost their wing secondly are grouped in the subclass pterygota. Main theories on the origin of insect flight are that wings developed from paranotal lobes are common in arthropods, including apterygotes (wingless insects). Rigid paranotal labes in some insects may be moved through deformation of the thorax by using indirect muscles attachments as seen in extent insects (Rasnitsyn, 1981).

   i) *Pleural-origin Theory:* Wings had their origin articulated flaps that extend from pleural region of the thorax (Trueman, 1990).

   ii) *Epicoxal Theory:* According to this theory, these tracheal gills started out exist of the respiratory system and over time were modified into locomotive purposes of wings (Jockusch, 2004).

3. **The Evolution of Pupal Stage of Insects:** Among the modern insect 2% belong to the group developed after some step of evolution. The 4th and 5th step in insect evolution provided insect with further opportunities to exploit their environment. They developed the capacity to retain their wing pads internally as imaginal dises. Hence they are called endopterygota. Since wing development is suppressed in the imature stage the wings must develop rapidly prior to emergence of the adult winged form. Presence of pupal stage permitted the suppression not only of wings but even of legs, eyes, mouth parts glands etc. The evolution of complete metamorphosis set in motion a remarkable two fold pattern of evolution. Larvae

were able to develop special structure of their own such as gills, prolegs on the abdomen, specialized mouth parts, glands etc. The major order of endopterygotes are coleoptera under all Beetle insect and Lepidoptera all butterflies and moth, Diptera order under the flies, mosquitoes and Hymenoptera all ants, wasps, honeybees insect.

4

Digestive System of Insect

The digestive system of insects comprises of two main sub-parts, one is the alimentary tract or also known as the digestive tract and the associated digestive glands. These digestive glands may be associated directly or indirectly such as, salivary glands, gastric, caeca (directly) and Malpighian tubules (indirect). If we have to describe an insect and its anatomy, or its histology, it is based on the orthopteroid plan, which is mostly shown by grasshoppers, crickets, and locusts, because their plan shows and describes the basic, primitive design from which other groups of insects have enolved and showed modifications and specialization. We can say that this orthoptera insect shows the general anatomical and histological features of a generalised insect.Today, we see that insects have invaded and adapted in almost all types of habitats and the major reason for their biological success is their ability to show digestion, utilization and their ability to eat an extremely vast diversity of food. It is due to this abilityof extremizm in food diversity that insects have undergone many specializations and modifications of the alimentary canal. What modifications a species of insects undergoes, depends upon the type of insects, depends upon the type of food it consumes. The diversity is also seen in the way of obtaining, storing, processing and absorbing the food. These differences may be structural or functional or both. It is not necessary that these differences must be present in different species only, those may be shown by different life stages of a single insect, by two sexes of a single species, etc. *e.g.*, the difference between the mouths parts of mosquito species.

ALIMENTARY CANAL

In insects, the alimentary canal consists of a tube-like structure, sometimes known as the digestive tube, which starts at mouth and ends at the anus. The mouth or oral cavity lies at the anterior end of this tube and anus lies at the posterior end of the digestive tube. Epithelial cell form a one-cell-thick layer of gut, these epithelial cells inturn lie on the non-cellular basement membrane known as the basal lamina. All the processes *viz.*, ingestion of food, its chewing, digestion and its absorption into the haemolymph and egestion are associated with their one-cell-thick gut. Alimentary canal comprises of three main parts, the foregut known as stomodaeum, the midgut called the mesenteron and the hindgut called the proctodeum. So, we see that stomodaeum is the anterior part of the alimentary canal; mesenteron is the middle part, while the proctodeum is the posterior part. The stomodaem and proctodaeum arise from the invaginations of the embryonic ectoderm and are lined by a hard, chitinous intima, which inturn is continuous with the cuticle of integument. We known that insects show moulting of the cuticle el integument, therefore during this process, both foregut, hindgut contents are also shed. The mesenteron is derived from the endoderm of the embryonic layers. Alimentary canal has also the longitudinal muscles and the circular muscles. The anterior part of the alimentary canal muscles are innervated by the stomatogastric system, and the posterior part muscles are innervated by nerves, which inturn arise from the posterior ganglion of the ventral nerve chain. Not only the longitudnal and circular muscles give support to the body, but also the trachea and tracheoles give support to gut. The another fact is that alimentary tube is shorter in carnivorous insects, which consume a high-protein diet, and is longer in herbivorous insects, which consume a high-carbohydrate food. The detailed description of the three main regions of the alimentary canal of insects is as under

Foregut or Stomodael

Morphological divisions are present in the foregut, and along with these divisions, the foregut serves the function of conducting food from pre-oral cavity to the mid gut of alimentary tube. At the base of the hypopharynx, the mouth lies within the pre-oral cavity or cibarium, formed of the mouth parts. It is in direct communication with the pharynx. The pharynx varies in various species of insects. Those insects which show a sucking mode of feeding, the cibarium and pharynx together form suctorial pump with dialator muscles. Pharynx then lead to the oesophagus, which mostly has a posterior enlargement called the crop. Crop stores food temporarily for sometime. Posterior to the crop, comes the gizzard or the proventriculus, it may be differently modified in different insects. It is absent in fluid-feeders, but is well- developed in orthopteroid insects *e.g.*, cockroach, grasshopper in which the intima of proventriculus is modified into six strong teeth for grinding food. The action of sieving or filtering is done by the spines of proventriculus, which partly fuse together to form a sieve like structure. The stomodael valve communicates the midgut with the proventricular region. Stomodael valve is modified and developed differently in different insects. The foregut comprises of impermeable intima due to which no absorption takes place here. But some salivary enzymes act on the food in crop and carry out digestion to some extent. But, it must be noted that foregut is not the major digestive region, it only consists of salivary enzymes and sometimes, the regurgitated enzymes of the midgut.

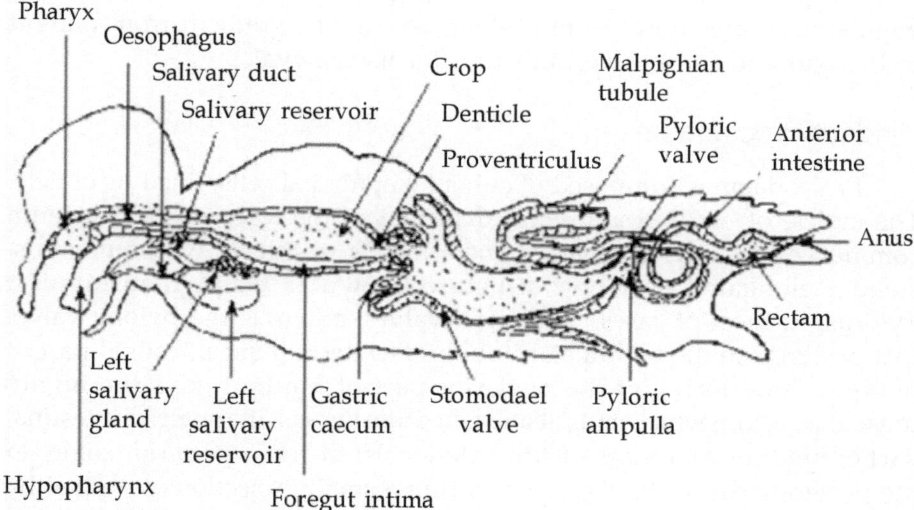

Alimentary canal of a generalized insect

Midgut / Mesenteric / Ventricular

Unlike the foregut, the midgut has no intima but is lined by peritrophic membrane in most of the insects, which is delicate and composed of chitin fibrils in a protein carbohydrate matrix. A group of diverticula, called the gastric caeca, are present posterior to the stomodael valve. In different insects, the number of diverticula varies differently.

Epithelium of the mesenteron comprises of columnar cells with microvilli, which are digestive and form a striated border of regenerative and endocrine cells. The plasma membrane of the basal part of digestive cells is infolded and a number of mitochondria are associated with these folds. The function of these cells is to synthesize enzymes for digestion and to absorb the digested food. The regenerative cells replace the actively functioning gut cells that degenerate due to holocrine secretion. The epithelial cells may be damaged due to hard food particles; therefore the bolus food gets surrounded by a peritrophic membrane. This membrane is permeable to enzymes prevents abrasions of the epithelial living of the midgut.

In the fluid-feeding insects, the gut is modified in such a way that excess of water is removed immediately to prevent excess dilution of haemolymph in order to provide a concentrated fluid to draw maximum nutrients *e.g.*, sap-sucking plant bugs. In other insects like aphids, a bladder is present between midgut and hindgut, which is closely bound to hind gut's anterior part and malpighian tubules by a basement membrane, water flows from it along an osmotic gradient from midgut to hindgut and from there to rectum for its removal.

Hindgut/Proctodaem

Proctodaem is composed of cuboidal epithelial cells, lined by cuticle. The cuticle of hindgut is thinner than that of the foregut. Proctodaeum commences with pylorus part of gut, which is associated with numerous, slender elongated excretory structure known as malpighian tubules. Pylorus, in most of cases, contains a valve known as the pyloric valve. Just posterior to the malpighian tubules lies the tubular intestinal part of hindgut (anteriorly), at the posterior part of hindgut, lays the highly muscular rectum which ends at anus. In some insects, the anterior intestinal part of hindgut comprises of still anterior ileum and a posterior colon. In the posterior part of hindgut, the rectum a small projections supplied by a rich supply of trachea. These projects are present in lumen of rectum and are excretory in nature, therefore, metabolically very active.

The recycling of the ingested food material enormously takes place in this region because the major processes like water reabsorption from waste food material (undigested food), ion reabsorption, phermone production, water conservation through cryptonephridial system, respiration in larvae of dragonflies and also the modification of symbiotic microorganisms occurs in the midgut. Therefore, we see from the above processes occuring in the hindgut, that it is the busiest factory of the alimentary canal of the insects.

SALIVARY GLANDS

In the insects, the salivary glands are associated with the labial segments of the mouth parts. But it does mean that only labial part contains glands. There may be glands associated with hypopharynx, maxillae or mandibles also. The salivary or labial glands are paired structures present ventrally in forgut in the head and thoracic region and may rarely extend to the abdominal region also. Salivary gland differs in shape, size and secretion according to the type of food taken by different insects. Salivary glands may be acinar or tubular. Acinar type of salivary glands is present in orthoptera and dictyoptera while as tubular type is shown in diptera, lepidoptera and hymenoptera. Coming to the acinar type of salivary glands, a tiny duct present on each acinus, communicate with other similar ducts from other acini, and they together form a lateral salivary duct, which inturn laterally merge at anterior part to form common salivary duct, this duct empties at the bases present between the hypopharynx and the labium. This region, where the salivary duct opens is called salivarium. Those insects which have sucking type of mouth parts, the salivarium forms a syringe like structure due to which saliva is injected into whatever is being pierced. In cockroaches, there are salivary reserviors, with which the lateral salivary ducts communicate. The secretion of salivary gland is a clear fluid. The function of salivary secretion has been briefed as under:

1. They serve as lubricant for mouth parts.
2. They act as a solvent for food.
3. They secrete silk in caterpillers, wasps, bees and related species.
4. In certain flies, the puparial cases are glued to the support (substrate) by these secretions.
5. Act as medium for digestive enzymes.
6. Act as medium for various acoagulins and agglutinins.
7. Secrete antimicrobial factors.
8. Produce toxins in some insects for their defence and paralysing the prey.

Saliva of insects contains mostly the invertase and amylase, and sometimes may contain the protease and lipase also. Aphids depend on plant sap for their nutrition, therefore nature has provided their saliva with a pectinase, which helps its mouth parts in penertrating the plant cells by dissolving the cementing material, the pectin, of their cells. Coming to the haematophagus (Haeme = blood, phagy = feeders) insects, their salivary secretion contains anticoagulat agents. In some insects like cockroaches and grasshoppers, the salivary secretion is under nervous regulation of both stomatogastric and suboesophagal ganglion, in Diptera, these secretions is under the control of some unknown neurohormone. The chemoreceptors present on head region of mouth parts, cause salivation by situmlating the nervous system when an insect sees or percepts the food.

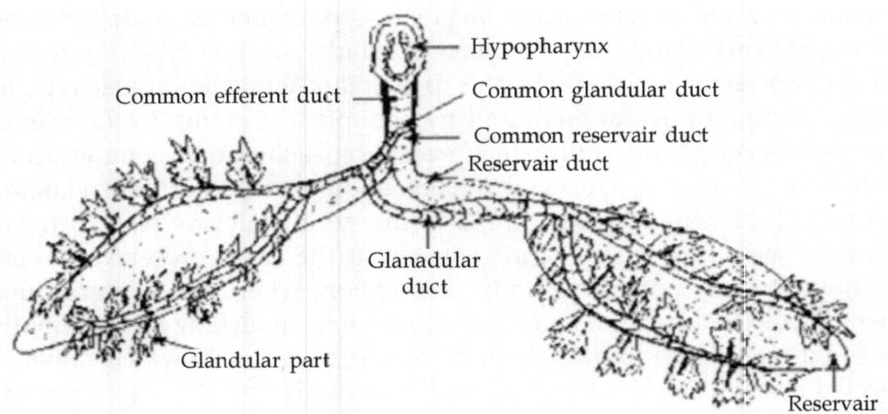

Salivary Glands of Cockroach

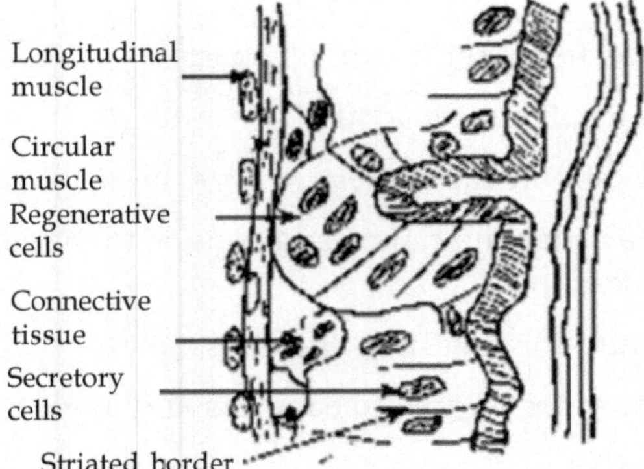

A section of midgut highly magnified

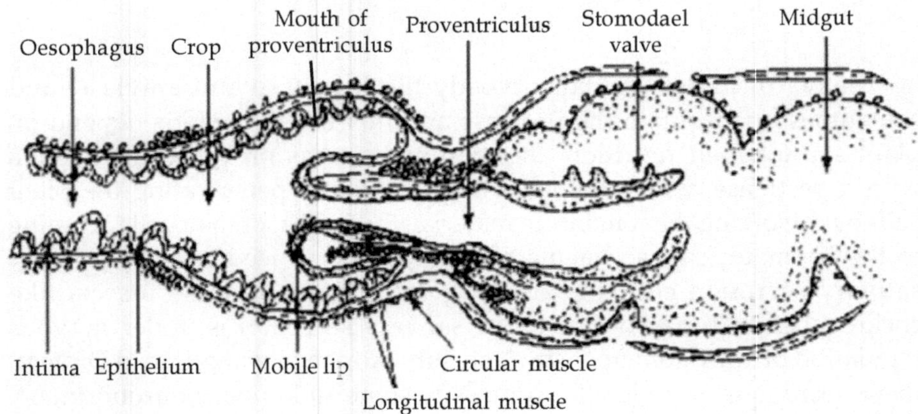

Longitudinal section of the proventriculus of bee

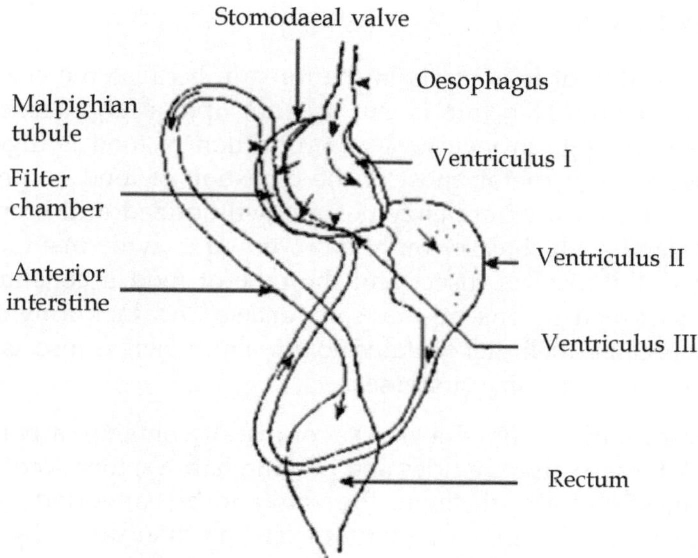

Filter chamber of pentatomidae bugs

Physiology of Digestion

Different insects have different type of mouth parts, depending upon the mode of their feeding. If we talk about the fluid-feeders, digestion begins when food is ingested through regurgitation of enzymes into the foregut, but most digestion occurs in the presence of all enzymes in the mid-gut. Those insects which possess chewing and biting mouth parts *e.g.*, cockroach, food is masticated not only in mouth but also in the gizzard. Through mastication of food in this way provides a greater surface area for enzymatic action on food, so that it can be converted into simple, absorbable material Depending upon the type of food taken by an insect, the type of digestive enzymes varies accordingly. It must be noted that all the enzymes, except the salivary enzymes, for digestion are secreted and synthesized in the epithelium of the mesenteron.

i) Extra-Intestinal Digestion

The digestion of food outside the alimentary canal before its ingestion is called the extra intestinal digestion. It is shown by predatory and parasitic insects, which inject saliva into prey and histolyses its contants. It is also shown by fluid-feeders. They drop their saliva on the sap before its ingestion. So, both these processes occurs outside the intestinal lumen, therefore, known as extra-intestinal digestion.

ii) Intestinal Digestion

The digestion of food starts in the foregut, because the enzymes of mid-gut are regurgitated into it. But it is not upto a large enxtent with some exceptions *e.g.*, in locust, large proportion of food is digested in crop. However, in general, most of the digestion of food occurs in the midgut of intestines, where enzymes are synthesized and secreted by epithelial glands, which carry out the process of enzyme-manufacturing according to the need of insect and the type of food it consumes. For instance, sap-feeding insects have invertase and lack amylase and proteinase because their diet contains no proteins, whereas insects feeding on protein diet synthesize proteases.

1. **Digestion of carbohydrates:** Diet of insects contains carbohydrates in the form of disaccharides and polysaccharides, therefore, in order to digest and absorb them, they have to be converted into their simpler units *i.e.*, monosaccharides. The mechanism of digestion of carbohydrates is shown as under:

   a) *Polysaccharides:* The major polysaccharide diet to be digested and absorbed by different insects involve starch, glycogen, chitin and cellulose

   Glycogen Amylase Glucose

   →————

   Starch/Amylose Amylase Maltose

   →————

   Both glycogen and starch are acted upon by amylase because this enzyme is specific to break the 1,4 a-glycosidic linkage in polysaccharides *i.e.*, starch and glycogen. When coming to the plant eating (phytophagus) insects, only a few contain cellulose enzyme to breakdown the cellulose. While most insects rather phytophagus insects don't contain/secrete the cellulase. In such insects, the cellulose ingested is either egested/excreted as such as they may contain symbiotic gut-microorganisms, which secrete cellulase. In the similar way, the chitin, ligocellulose and hemicellulose are digested and acted upon by chitinase ligocellulase and hemicellulase respectively.

   b) *Disaccharides:* Food of insects, which feed upon the carbohydrates, contains disaccharides in the form of sucrose, maltose, cellulose, trehalose, lactose and melibiose. All these disaccharides contain a glucose residue linked to a second sugar residue either by a α-linkage or a β-linkage.

During hydrolysis of these disaccharide units, water plays a major role as under;

⇒ Maltose + H_2O	Maltase	2 Glucose
⇒ Sucrose + H_2O	Sucrase	+ (fructose in Sucrose)
	Glucose	
⇒ Lactose + H_2O	Lactase	+ (Galactose in lactose)
⇒ Cellobiose + H_2O	Cellobiase	+ (Glucose in cellobiose)
⇒ Melibiose + H_2O	Melibiase	+ (Galactose in melibiose)
⇒ Trehalose + H_2O	Trehalase	+ (Glucose in Trehalose).

So, we can see that from the above graphic representation, during the hydrolysis of disaccharides in insects, every diasaccharide unit is broken down into glucose and an additional sugar in some units. Each hydrolysis reaction is catalysed by a specific enzyme.

2. **Digestion of lipids:** Those insects which consume fat containing food particles secrete lipases. The lipase enzymes hydrolyse the fat into fatty acids and glycerol. Some species of moth, which feed on the wax of bee-hive, are able to digest the wax due to presence of lipases, *e.g., Galleria*, commonly known as the wax moth not only contains lipase, but, with the help of symbiotic bacteria, it also contains lecithinase and cholinesterase. The digestive processes in insects include many factors, which vary from species to species and may also vary among the different regions of gut of the same insect. These factors include, oxidation- reduction potential, temperature, buffering capacity and pH of the midgut, which is usually 6-8 in insects.

iii) Absorption of the Digested Food

From the above discussion, we came to know that in insects, major abortion of food occurs in midgut. In the hindgut, only reabsorption occurs and no absorption occurs in stomodaem (foregut). In addition, no phagocytosis is shown by insects; therefore food is absorbed in the form of solution.

The digestion of food materials in insects is affected by four major factors, first is the microvilli, which increase the surface area for absorption, second is the permeability differences in different regions of gut, the presence of the counter current, because the substances move from the region of their higher conc. to the region of their lower conc. Absorption may be passive, when diffusion takes place from lumen of gut, where conc. is higher the epithelium of the gut where conc. is lower.

Absorption is active, when some metabolic energy is to be consumed for the movement of the molecules against their concentration gradient.

Carbohydrates are absorbed from gut lumen to the haemolymph, down the concentration gradient. Absorption occurs by facilitated diffusion also, where glucose and fructose are converted to trehalose in the fat body. The some insects are able to absorb disacharides as such. Proteins are first converted to amino acids and then absorbed in the midgut caeca. In hindgut, also, some amino acids are from urine. The characteristic feature of insects is that they store high quantities of amino acid reserves in their haemolymph, therefore they have to actively absorb amino acids against the concentration gradients, but exceptions are there, where insects absorb proteins and peptide fragments as such, it is seen in blood-feeding *rhodnius*, which absorbs haemoglobin as such. But, this type of absorption varies in different insects and that depends upon the composition of both food and haemolymph.

Lipids are also, sometimes, absorbed as such cholesterol is estermised by insect-midgut, while as wase is phosphorylated before absorption. But in some hymenoptera adults, lipids are absorbed in the hindgut. Water is absorbed both in the mid-gut and hindgut, by diffusion or by active transport. Because, insects precisely maintain their salt-water balance, therefore, the mode of water absorption depends upon the need of the insect. Filter chamber in some insects is a modification of the anterior midgut, which in combination with the malpighian tubules, remove excess water. Inorganic ions are absorbed in midgut and reabsorbed in the hindgut *i.e.*, rectum. There are special cells in the gut lumen which absorb the essential ions like Na+, K+ and Cl–. The Na+ pump inturn plays a major role in the absorption of other ions. They also create a water gradient between the lumen and epithelium cells and then from cells to haemolyph.

REGULATION OF THE DIGESTIVE SYSTEM

Digestive system regulation in insects is to regulate the control on food movement along the alimentary tube, control on secretion of enzymes and control of absorptions. It is partly regulated by the stomatogastric nervous system. Ingestion of food is aided by mouth parts such as cibarium and pharynx and temporarily stored in crop. Then digestion of this food is completed in midgut, by gradually allowing the food to pass from crop to the hindgut by the stomodael valve. The studies have shown that movement of food from crop to midgut is under the frontal ganglion of brain. Where a food goes after being ingested depends upon the type of food, it is called switch mechanism and is under the control of frontal ganglion.

Control on the rate of crop emptying has been throughly studied in *periplante-americana*, where the conc. of the food is inversely proportional to its passage. Some osmotic receptors have also been found in cockroach pharynx. Secretion of enzymes is controlled either by scretogogue or hormonal response. In the former one, the response to food is immediate while as in the latter, the control is related to developmental and environmental factors.

Absorption is mainly regulated by the availability and release of food from crop. The movement of food may be neural or they may be myogenic, but in some insects, these movements have been found under the hormonal stimuli also.

FEEDING BEHAVIOUR OF INSECTS

Types of Feeding what an insect consumes as food is related to the fact that where it lives. On the various modes of nutrient derivation, the insects may be grouped as under:

a) *Herbivores:* These are also known as phytophagous and plants are the main source of food for them. Among the insects, the herbivores constitute the largest group. These include grasshoppers, locusts, bugs, moths, butterflies, termites, fruit flies, etc. Some insects feed on the plants of a particular genus, these are called monophagous, while as others feed on the variety of plants called polyphagous insects. Different insects feed on different parts of a plant, but a particular insect feeds on a particular part of a plant *e.g.*, leaf rollers, root worms, stem borers, etc.

b) *Detritivores:* These insects are the ecologically most important members of earth, because they feed on the dead and decaying organic matter. Detritivores are mostly found in forests. *E.g.*, dung beetles, carrion beetles, etc.

c) *Microphagous:* Those microorganisms which are associated with decaying organic material are consumed by some insects, such insects are called microphagous, *e.g.*, maggots of the Drosophilla.

d) *Carnivores:* These insects feed on flesh or body parts or body tissues of living animals. Carnivores may be predators, parasites or parasitoids.

   1. *Predatory insects* eat other living organisms, mostly insects, by capturing them. These include dragonflies, predatory wasps, laccwings, etc. The predatory insects are adapted in such a way that they have an accurate power of vision, swift and fast speed,

and differently modified body parts to catch their prey in an accurate manner. Some predatory bugs are so big (Giant bugs) that they feed on the water snails. Some form traps to deceive their prey in order to catch it.

2. *Parasites* are those which live in the body of their host called endoparasites or on the body of their host, called the exoparasites. These do not kill their host but derive their nutrition from it, and some vertebrate-parasite insects spread diseases in them. *E.g.*, human louse lives entirely on human body, therefore is an exoparasite and drives its nutrition from its host, rat-flea is a parasite of rats, etc. Parasites living on a host are always host specific, and may be permanent or time-being for a host.
3. *Parasitoids* are those insects whose larval forms feed on the body of other insects, while the adults are free living some parasitoids show a high degree of host specificity, and some are economically very important for the biological control of pests due to their property of host specificity.
4. *Mycetophagus* insects feed on fungi.
5. *Inquilines* are those insects which live in the shelters of other insects and feed there without causing the harm to the permanent owner of that shelter. *e.g.*, ants of different species share common colony.
6. *Trophallaxis* is mostly seen in social insects. The adults feed their larvae from their mouth and in turn get secretion of salivary glands from larva. This is a mutual mode of feeding. Termites have rectal symbionts and pass them on to next generation by anal trophallaxis.

5

Circulatory System of Insect

Insects, belonging to the phyllum-arthropoda, have an open circulatory system. An open circulatory system is that in which blood bathes the organs rather internal organs directly in the haemocoel, the coelom of the insects, which actually is not a true coelom. Almost entire haemocoel of insects is formed from the epineural sinus. The haemocoel is divided into the perineural sinus and perivisceral sinus by the dorsal and ventral diaphragm respectively. The characteristic of insects is the presence of dorsal vessel, which comprises of anterior aorta and a posterior heart and is the only conducting tube present in insects. The blood of insects mostly does not perform gaseous exchange, therefore it does not contain haemoglobin, and blood in insects is called the haemolymph, which is red-coloured. So in order to study the circulatory system of insects, the major focus must be on the dorsal vessel, which has been described as follows, along with its accessory structures.

DORSAL VESSEL AND ACCESSORY PULSATILE PARTS

i) Dorsal Vessel

As already said, in insects, dorsal vessel is the major circulatory organ in the process of circulation. It is situated at the dorsal side, at the middle line of the insect body extending from posterior body region to head. The vessel wall constitutes, circular, oblique, helical, semicircular or longitudnal muscles fibres. But mainly there are circular muscle fibres. This dorsal vessel is highly contractile and is generally a straight, simple tube, but may be some bulging thickings in some cases connective tissue covers the vessel on outside. The respiratory tracheoles are associated with it. The dorsal vessel is suspended in the body by the fibres from alimentary canal, integument and other structures.

1. **Heart:** Dorsal vessel is divided into a posterior heart and an anterior aorta. Heart is closed posteriorly, and is mostly restricted to the abdomenal region. It has a number of openings, called ostia, the

ostia are lateral. Those ostia which allow haemolymph to enter the heart are called incurrent ostia while those through which blood leaves heart are called excurrent ostia.

a) *The incurrent ostia:* There are usually three pairs in thoracic region and nine pairs in abdominal region. These ostia are vertical like a slit in the heart wall and act as its openings. But it is not necessary that these pairs are accurately numbered in all insects. The number of pairs of ostia may deviate in many species, *e.g.*, it is three pairs in houseflies but five pairs in wasps. Every ostium has a valve which determines the flow of blood in and out of heart at diastole and systole respectively.

b) *Excurrent ostia:* The excurrent ostia show movements opposite to that of incurrent ostia. They expand at systole while they contract at diastolic movement of heart. The blood from the excurrent ostia flow in perivisceral and pericardial sinuses from heart. These ostia are paired, and do not possess valve. These are ventro-laterally present in thoracic and abdominal region *e.g., Lepisma, locusta,* etc. The heart in insects is not chamberus and its lumen is continuous throughout its length, but due to constrictions between the ostia, it looks like a chambered heart *e.g.,* cockroach. Heart is attached to the body wall on both sides by the special alary muscles, which hold and gives support to heart inside the body cavity of the insect. Those insects which lack excurrent ostia segmental vessels with valves are present which allow the blood to move out of heart.

2. **Aorta:** Aorta is a tube-like structure/organ, which bears no ostia, but in some cases, its lumen is divided by septa or diverticula. The diverticula are inturn connected to heart and dorsal vessel. From heart, it extends anteriorly and opens behind the brain or sometimes beneath the brain.

ii) Accessory Organs

The dorsal blood vessel alone does not regulate the flow of blood throughout the body of insect. Haemocoel contains some accessory organs also, which are pulsatile and add to regulation of circulation in insects. In moths, the dorsal vessel coils and goes upto the thorax to form pulasatile organs while in most of the insects, the mesothorax and metathorax bear the pulsatile accessory organs, which supply blood to the wings. When we talk about cockroaches and grasshoppers, we definetly remember the ampulla at the bases of antennae. What is the function of this ampulla? The ampulla is continuous anteriorly with the lumen of antenna. At the posterior side or base of the ampulla, a valve is present, and from base of antennae, through this valvular aperture, the base of ampulla is in communication with dorsal vessel. When ampulla contracts, the blood moves from it to the haemocoel and when ampulla expands, the valve opens and blood flows into the antennae.

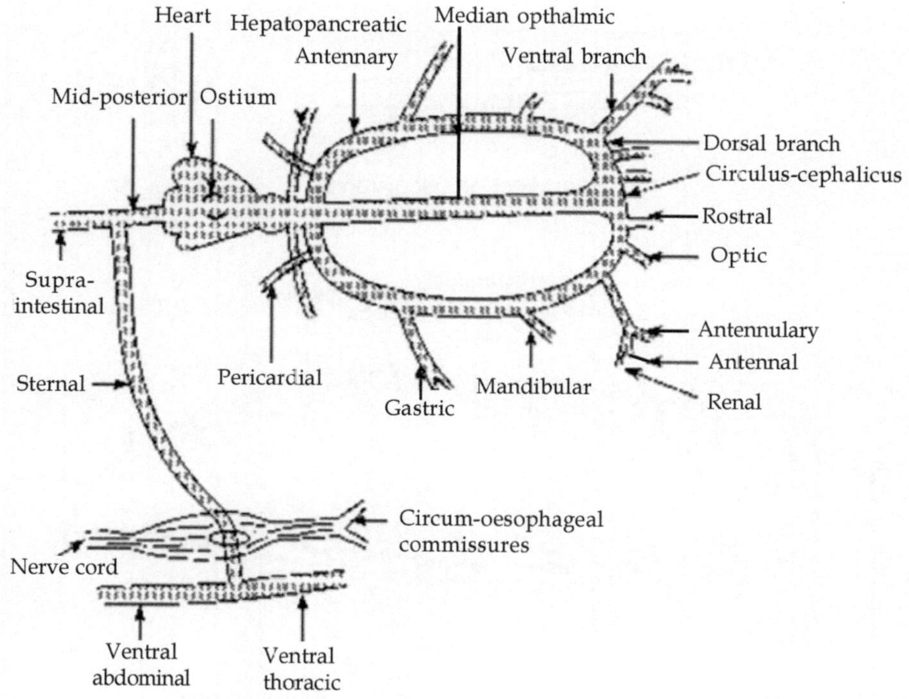

Blood Vascular System of Insect (Palaemon)

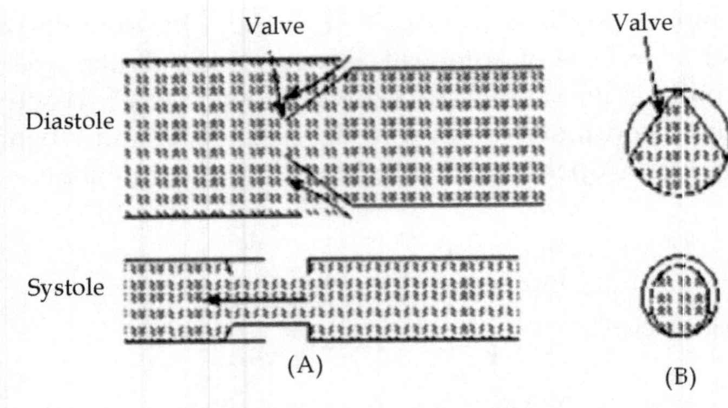

Heart section ostial valves

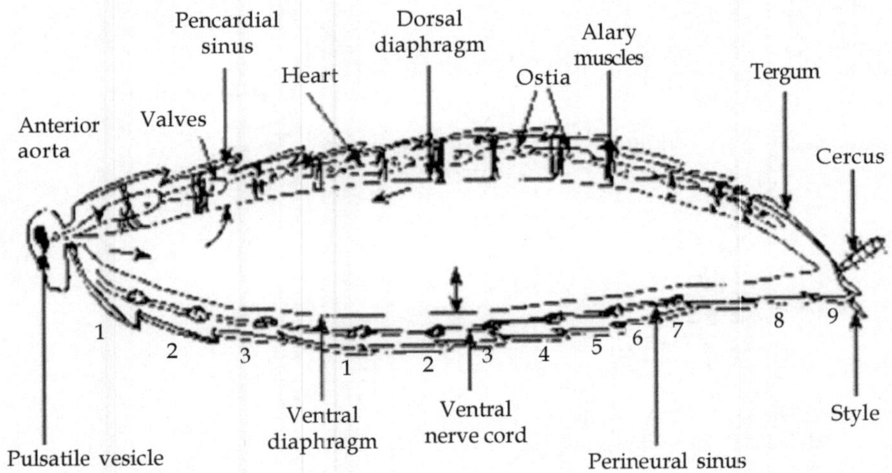

Blood Vascular System of Insect

iii) Diaphragms, Sinuses and Alary Muscles

Sinuses are the cavities into which the haemocoel of insects is divided by the fibromuscular septa known as the diaphragms. The diaphragm is dorsal and ventral, and dorsoventral movements of both these diaphragms determine and regulate the circulation of blood in the haemocoel. Dorsal diaphragm is also known as pericardial septum; it encloses alary muscles between its layers, which hold the heart with the body wall. The pericardial sinus, *i.e.*, the diaphragm around heart is separated from periviseral sinus, *i.e.*, the diaphragm around viseral organs, by the dorsal diaphragm. It is perforated laterally. Ventral diaphragm divides the periviseral sinus from the perineural sinus *i.e.*, the diaphragm around the nerve cord. Ventral diaphragm is located just above the nerve cord and is not present in all insects.

Alary muscles are the fibromuscular strands and are entirely associated with heart to suspend and support it in the body wall. The alary muscles from both sides of the body wall meet at the midline of the body.

iv) Mechanism and Course of Circulation

The bloods enter the heart by in current ostia and leave it through the excurrent ostia. The valves of incurrent ostia prevent the backward flow of blood. In the head region, blood returns to the general circulation. Blood circulates throughout the haemocoel and other appendages by the movements and contractions of alary muscles, dorsal diaphragm, ventral diaphragm and other accessory pulsatile organs. Course of blood is as under.

Blood → Ventral diaphragm → Perineural sinus → Openings of dorsal diaphragm → Periviseral sinus → Pericardial sinus → Heart.

HEART BEAT- ITS REGULATION

Heartbeat of insects consists of two phases, the systole and the diastole. During systole the heart contracts and undergoes a decrease in volume, this contraction spreads as a wave from posterior end to forward direction. During diastole, the heart muscles relax by the contraction of the alary muscles; it is rather a relaxation phase of heart muscles. Therefore, the systole and diastole together constitute the heartbeat cycle and diastosis is a resting phasbetween the two successive beats. Regulation If we talk about the regulation of heartbeat, then it is to be noted that in insects nervous system plays, no role, except a few, like cockroaches in which nerves are supplied to heart from corpora cardiaca and ganglia present in nerves, which are motor in function. Then the question arises how an insect heartbeat works, if it is not under the control of brain? It is logical and simple, that insects possess a myogenic heart, with no pacemaker. Heart beat in insects is regulated by cardio accelerator neuro peptide and proctolin. These are the myotropins, which act on heart.

Therefore, this regulation is neurohormonal. In the larvae of some insects, the heartbeat alternately increases forwardly and backwardly, but during adult period, it increase only in forward direction. It is because of this reason that wings expand and unfold during adult stage. Responses to stimuli, either external or internal have also been found in heartbeat of some insects, *e.g.*, callipora. Heart in insects is also affected by various factors like.

(1) Temperature, (2) Activity, (3) Metabolic rate, (4) Insecticides, (5) Developmental stage, (6) Drugs, etc. The hydrostatic shape of body in some soft bodied insects is maintained by blood pressure of haemolymph. Heartbeat varies in different species of insects or even among the different life-cycle stage of a single insect. The regulation of heartbeat is therefore maintained by dorsal blood vessel and acessory pulsatile organs, though they both act Independently of each other.

INSECT BLOOD

Blood of insects is known as the haemolymph, which is a clear colourless fluid in most insects, but may be green, yellow or red due to the presence of some pigments in some insects. Some insects like chirononus larva, gasterophilus, etc., contain haemoglobin. Blood or haemolymph constitutes 8-40% of total body weight of an insect, pH is usually 6-7 or in a few insects may be 7-7.5, having specific gravity of 1.015-1.060, but during moulting it may rise.

Composition of Haemolymph

Haemolymph constitutes 81-88% of water, while other few insects may have water content of 40-42%.

1. Inorganic constituents include sodium, K^+, Ca^{2+}, S, Mg^{2+}, P, Cl^- and HCO_3^- ions. The presence of ions in insects also depends on their feeding habit. If insect is phytophagus, Mg is the major cation. While as carnivorous insects have high amount of Na as compaired to Mg^{2+} and K^+.
2. The metabolic wastes are in the form of uric acid present in a high conc. in the haemolymph, and are excreted by the malpighian tubules. Other nitrogenous wastes are urea, allantoic acid and ammonia (in aquatic insects).
3. For the translocation of proteins, for identification tags in cellular recognition and in the cold stress/hibernation metabolism, carbohydrates are the main role players. Trehalose, a dissacharide compound contains double energy than that of a glucose molecule, and it acts as a major blood sugar of insects. In addition to this, other sugars like glucose, fructose, ribose, etc., are also found. The maintenance of blood sugar is done by hyperglycemico decapeptide and the hypoglycemic, both of which are the neurohormones, present in the haemolymph. The scout cells take place and function of WBC's which act against the foreign and toxic bodies in the insects. In cold stress, many antifreezers or cytoproctant chemicals, like glycerol and sorbitol, are present in insect haemolymph.

4. Organic compounds include lactate, pyruvate, fumrate, succinate and many other enzymes and co-enzymes of respiration. These also help in maintaining the blood cation level.

5. Lipoproteins or lipophorins are the lipids present in haemolymph. These also help in transport of fats in insects from gut to various tissues. The eggs of insects also contain recognition glycolipi proteins, one of which is vitellogenin, which transport lipids into egg for the developing embryo.

6. Proteins are present in many forms in insects. Proteins are present in plasma membrane, in cell constitutents *e.g.*, enzymes are also proteins, antibactericidal proteins, carrier and channel proteins, juvenile hormones esterase, lysozyme and phenoloxidase all are the proteins. During larval development a protein called calliphorin, functions as a store for nutrients.

7. Different amino acids are also present in haemolymph. Free amino acids are present in high concentration. These amino acids are very important for the metabolic activities of insect body, which are either synthesised in body or taken from outside in diet various amino acids play a major role in metamorphosis, cuticle of insects contain tyrosine for its sclerotization.

8. Though blood of the insects is clear, colourless, fluid called haemolymph, but some insects have some respiratory and non-respiratory pigments in their blood, due to which their blood gives a characteristic colour *viz.*, green, yellow, blue or brown. The haemolymph having blue colour is called haemocyanin *e.g.*, in cockroach. The colour of blood may very in different species or sometimes in different stages of life-cycle of a same insect *e.g.*, the difference in blood colour of adult cockroach and cockroach nymph. Some insects like chronomous contain haemoglobin as respiratory pigment. The colour of respiratory pigment may sometime depend upon the meals taken by the insect. *E.g.*, herbivorous plant sap feeders bear green coloured pigments while as blood sucking bugs show red pigments.

9. Our blood contains haemoglobin that is why our blood contains a high amount of O_2. But in insects, they are no such a pigment; there is very low concentration of O_2 and CO_2. But in chronomous larvae, blood contains high conc. of O_2, because of the presence of haemoglobin.

Haemocytes

Haemocytes are the cells present in the haemolymph. Haemocytes of insects has been originated from mesodermal layer and are unpigmented. The haemocytes do not enter the dorsal blood vessel and its accessory pedsatile organs. Mesoderm of embryo contains prohaemocytes from which haemocytes develop. These prehaemocytes are present in haemocytopoietic organs. In exopterygota, the haemocytopoietic organs are present in both larval and adult stages as well as in embryonic stages. Number of haemocytes varies with environmental factors and diet of insects. There are a number of haemocytes present in haemolymph, which are morphologically different and vary from 10-167000 ml, and depends upon the species, developmental stage, physiological state, etc.

a) *Prohaemocytes:* These are small rounded cells with relatively large nuclei and basophillic cytoplasm, also give rise to other haemocytes.

b) *Coagulocytes:* These haemocytes are granular cystocytes having small nucleus and a pale, hyaline cytoplasm with black granules.

c) *Plasmatocytes:* These are most abundant, variable in shape and form, phagocytic with basophillic cytoplasm.

d) *Granular phagocytes:* These are phagocytic with acidophillic granules in cytoplasm.

e) *Oenocytoids:* They are usually large, thick, basophillic cytoplasm. They also have the canaliculi.

f) *Spherule cells:* These haemocytes are round, oval are filled with large, they are non- refrigent, usually some acidophillic inclusions are present. In additior., some insects contain small spheroidocytes, having refringent fat droplets, these are calle *adipohaemocytes*.

Functions of Haemolymph Haemocytes provide protection against foreign bodies, pathogens and toxins. These act alone or act in association with the haemolymph following are the main functions of haemocyte:

1. Transport and storage is the main function of haemolymph to transport nutrients in every part of body. They are transports of different hormones, gases and metabolic wastes.

2. Protection against pathogens, infectious bacteria and cell debries is done by phagocytotic haemocytes.

3. If any kind of bacteria enter they haemolymph of an insects they are removed by haemocytes, by forming a nodule like covering

around them, wherein they are trapped, there nodule forming haemocytes are called plasmatocytes. In this way a large number of bacteria are removed from the haemolymph.

4. Other haemocytes, called granulocytes act when the foreign bodies are enough bigger to be phagocytosed or nodulated. The granulocytes burst and their inner material sticks to the foreign body. Meanwhile, other granulocytes also attack the foreign particle and release a factor known as haemocytic recognition factor, which attracts plasmatocytes to cover and to get flattened all over the foreign body, until it no long remains non-self and foreign for the haemolymph.
5. The Certain haemocytes and haemolymph also carry out the function of detoxification.
6. It,s the site of injury, plasmatocytes and spherule cells, accumulate to heal the wound, granulocytes promote coagulation and form sheets of cells on wound for healing. The neorn out cell debries are phagocytosed by plasmatocytes and spherule cells.
7. Haemocytes precipitate the plasma at the site of mound in order to prevent loss of haemolymph. This is done by special type of haemocytes, called granulocytes which carry out the function of plugging a wound.
8. Haemolymph lubricates the internal organs in order to allow their movement with respect to each other.
9. Haemolymph also provides turgidity by its hydraulic pressure on the body wall. This hydraulic force helps the newly born young to emerge out of egg by forming hydrostatic intrusions. Moulting and ecolysis also requires the hydraulic force.
10. Haemolymph also keeps the temperature same all over the body by transferring heat from one part of the body to another.
11. Unlike humans, insects do not have an immense system or antigen-antibody system. But still haemolymph protects them from the invasion of bacteria, viruses and other foreign bodies. The insect immune system simply helps the insect to resist the foreign invasion. The immune factors present in haemolymph include inducible and non-inducible. Inducible factors include antibacterial proteins called cecropin and lysozymes. Non-inducible factors do not require RNA and protein synthesis and include lectins called haemaglutinins and phemyloxidases. Phenyloxidase remains inactive in absence of foreign body, but once this system is activated it produces a cascade of reaction to kill the invader.

12. Haemolymphy along with haemocytes carryout the function of secretion and formation of other tissues. They form the basement of epithelium and sheath cover of muscle fibres. Prothoracic glands are activated by certain special haemocytes to induce moulting.

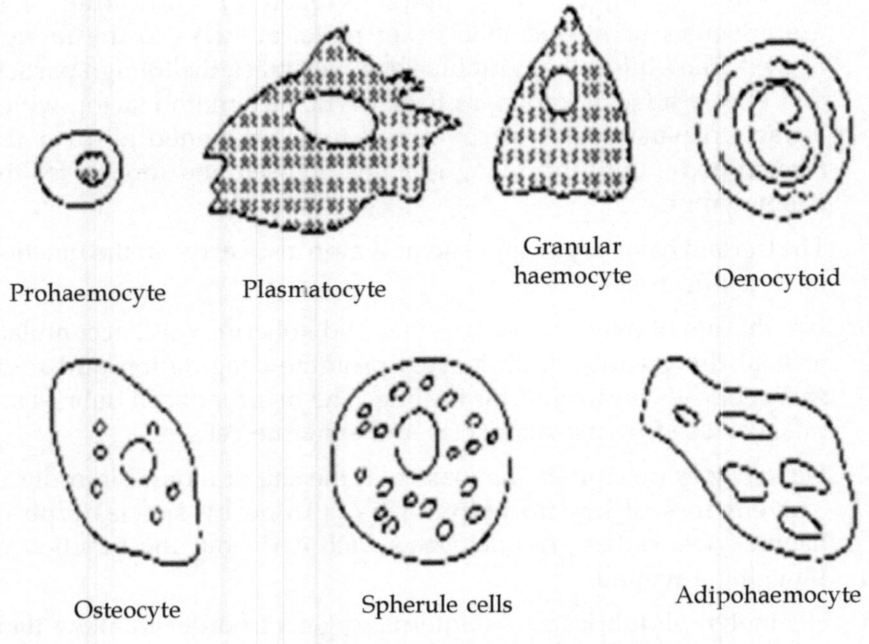

Types of Haemocytes of Insect

6

Excretory System of Insect

Excretory system of an organism maintains its homeostasis *i.e.*, a constant internal environment. Here in insects, the blood is haemolymph therefore it determines the excretion of an insect's body. Excretion also eliminates the nitrogenous wastes and maintains the salt- water balance and also regulates the uniformily of haemolymph. The characteristic feature of insects is the presence of malpighian tubules to eliminate the metabolic wastes, whereas rectum walls carry out the function of reabsorption of salt and water. A minimum amount of water is excreted with nitrogen waste in order to conserve the water in body.

i) Malpighian Tubules

Malpighian tubules are the blind-at-one end-tubes, attached to the alimentary canal at the junction between the midgut and hindgut, inside the haemocoel. Each tubule is slender, long and blind tube and may directly open in the mid-gut or hind gut or into amupullar structure. These tubules are convoluted and hand freely in the haemocoel. The no. of malpighian tubules varies in different species and some insects *e.g.*, aphids, even lack the malpighian tubules. The malpighian tubes are arranged in the posterior part of alimentary canal in two ways:

1. **Gymnonepuridial arrangement:** In this type, the distal end of tubules lies freely in haemocoel. It can be further of two types :
   a) *Orthopteron type:* This type of malpighian tubules possess same type of tissues and are only secretory in nature.
   b) *Hemipteran type:* The basal absorptive part of malpighian tubules differs from the distal secretory part histologically.
2. **Cryptonephridial arrangement:** The distal end of tubules are embeded in the tissues surrounding the rectum. This arrangement of tubules ensures the maximum reabsorption in rectum. It is further divided into two types:
   a) *Coleopteron type:* The malpighian tubules are alike and are secretory in nature.

b) *Lepidopteran type:* The basal region of malpighian tubules differs from the distalend. Histology of Malpighian Tubules are associated with the serpentine movement producing tubules, these propel effectively the contents of lumen towards posterior opening of the gut. It also ensures the thorough mixing of the contents of digestion and exposure of tubules to more haemolymph. These tubules are one cell thick and are traversed by a network of tracheoles, these cells lie on a basement membrane. Cytoplasm is colourless, filled with pigmented inclusions and crystals. The distal end cells are produced into small outgrowths which anastomose together and carry out the secretory function, while the proximal end cells bear a brush border. These brush-bordered cells are concerned with active absorption. The cells of malpighian tubules are rich in mitochondria, because they are the active organs of excretion needing maximum amount of energy.

ii) Nephrocytes

The nephrocytes convert the waste materials into a form that may easily enter the metabolic pathway. Nephrocytes regulate protein metabolism and heart beat. They occur in groups in several parts of the body or are single. Their size differs in different species. *E.g.*, diptaranes larvae, they are large, while as in others, they are small and multinucleated. They are in close association with pericardium, hence are also pericardial cells.

iii) Rectum

The lost part of alimentary canal lying in the hind gut is the muscular rectum, which bears seetal pads in its lumen and carries out the function of reabsorption. It is highly tracheolated. In some aquatic insects, ammonia is secreted directly into the lumen.

iv) Other Excretory Organs

Some insects *e.g.*, springtails do not have malpighian tubules; uric acid is excreted by other organs like:

1. **Labial glands:** Labial glands consist of an upper sacchule followed by a called labyrinth and have a gland opening.
2. **Utricular glands:** Uric acid, in some insects is stored in the utricular glands, present in males called male accessory glands and is poured out over the spermatheca during copulation. Females also store their own uric acid and during copulation both parents pass a part of their uric acid to embryo, which provides an alternate source of

nitrogen to them. *E.g.*, In the members of cockroach family. In the embryo, the myecetocytes of fat body contain microorganisms which produce uricase enzyme to act on uric acid, which hydrolyses to produce nutritional nitrogen source.

3. **Fat Body:** Urate cells of fat body, in oriental cockroaches, store uric acid. This is a store for nitrogen.
4. **Other tissues:** Uric acid also accumulates in epidermis of some insects and at every moult, it is removed e.g., rhodinus. During pupal stage of butterflies, uric acid is stored in scales of wings to produce new tissues and also to supply adenine for nucleoprotein synthesis.

NITROGENOUS EXCRETION

1. **Excretory Products:** As a result of protein, amino acid and nucleic acid metabolism, various types of nitrogenous wastes and products are accumulated in the haemolymph, which are of no use to an insect and sometimes may be toxic, therefore it must immediately either be removed or stored in an inert form, until they can be used next or be excreted. Nitrogenous wastes include ammonia, urea, uric acid, allantoin, allantoic acid, amino acids and even proteins. The terrestrial insects are vericotelic, aquatic are ammonolelic uric acid does not need much water for its elimination; therefore it is the main excretory product of most of the insects. The halsitat of an insect greatly determines its excretory product. Butterflies and moths excrete allantoic acid while as red cotton bug excretes allantion urea is present in a very small quantity in insect excretions. *Glossina,* the tse-tse fly, sucks the blood of humans, which contains arginine and histidine. Both these aminoacids are excreted unchanged after absorption.
2. **Mechanism of Excretion:** Malpighian tubules are highly permeable to small molecules; they therefore, filter materials in excess from haemolymph. They enter the tubules either by simple diffusion or by active transport by K+. Toxic substances are excreted or removed while the useful substances are reabsorbed in distal part of tubules or in the rectum. Waste is excreted as faeces from anus.
3. **Salt-Water Balance:** It is also known as the osmoregulation. Due to different environmental conditions, different salt and water stresses are posed on insects. Terristrial insect's loose water by constant evaporation and required water and salt is supplied in diet. Those insects which feed on plant sap, the faeces are watery, while those which feed on low water content food excrete faeces in the form of dry powdry pellets. The excess water and salt is removed and the required amount of salt is reabsorbed in the rectum, where active transport does not occurs always.

4. **Dietary Problems and Excretion:** Insect's definetly derive energy from the food they consume. But to maintain homeostasis, a right amount of salts and ions must be reabsorbed, so that no dietary problems may arise, between the conc. of food and the haemolymph. In addition to this, herbivores insects face the problem of phytotoxic chemicals. These chemicals are continuously removed from the haemolymph by excretory malpighian tubules but some insects retain them for their use *e.g.*, grasshoppers.
5. **Control of Diuresis and gut Motility:** The production of urine, called diuresis, is controlled by diuretic or antidiuretic hormones, secreted by corpus cardiacum of brain. Dipeptide in locusts and ADH in house crickets, affect the malpighian tubules. A chloride transport stimulating hormone regulates water and ion balance in locusts. Proctolin in cockroaches acts as an excitatory neurotransmittor, having myotropic effect on visceral muscles of hindgut.

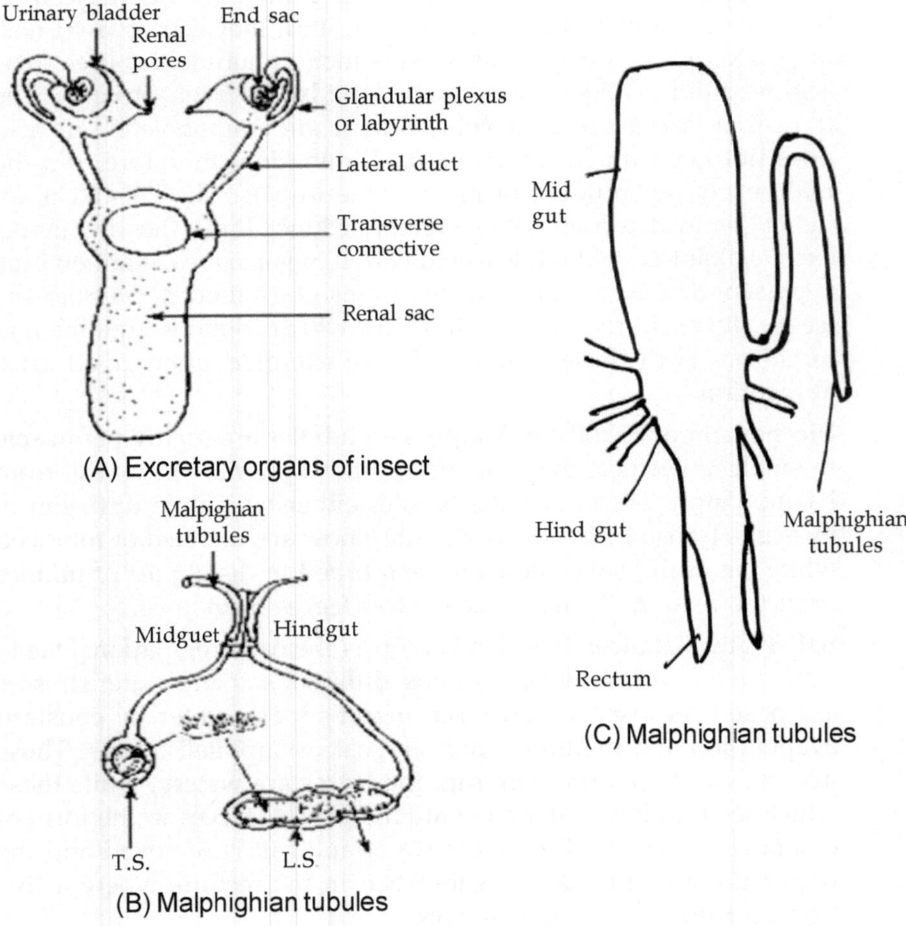

(A) Excretary organs of insect

(B) Malphighian tubules

(C) Malphighian tubules

7

Reproductive System of Insect

Insects show sexual Dimorphism *i.e.,* the sexes are usually separate and are dioecious. Male reproductive system produces sperms, stores them and delivers them *i.e.,* spermatozoa. It also produces the seminal fluid, which provides a suitable environment to the sperms and nourishes them. Male reproductive system also situmilates are the female for oviposition. The female reproductive system produces and stores eggs provide nutrition to them for embryonic development, recieves and stores the sperms, is a site of fertilization. Female reproductive system also deposite eggs and may provide additional protection to the embryos. The reproductive system of both male and female insect's bearis a gononpore, which is an opening of a pair of gonads. Both the gonads open into the gonopore by a median duct. For the secondry sexual purposes, *e.g.,* formation of spermatophore and egg-case or ootheca, some accessory glands are associated with the reproductive system.

MALE REPRODUCTIVE SYSTEM

Male reproductive system in insects is located in the abdomen and typically consists of paired testes connected by ducts, which lead into the copulatory or intromittent organ (adeagus or penis). The ducts of the male reproductive system also bear accessory glands.

i) Testes

Testes are the paired structures lying in the abdominal region. Each testis has a number of sperm tubus also called testicular follicles, *e.g.,* there is only one follicle in certain beetle's two follicles in lice and more than 100 in grass-hoppers. The follicle wall is made of epithelial cells lying on a basement membrane. The function of these follicles is to derive nutrients from haemolymph for the nourishment of the developing germ cells. In some insects like moth or locust, the two testes are close together and sometimes fuse. The testicular follicles bear germarium at their distal end, which contains spermatogonia that undergo the process of spermatogenesis to develop into sperms stage. This process occurs at last instar or pupal stage, and in some species, it continues in the adult stage.

a) Large apical cell or group of cells are contained in a follicle which nourish the spermatogonia, which are associated with other cells forming a cyst. There are three zones of development in the germarium and represent the different stages of spermatogenesis: The growth zone or spermatocyte zone, where the deploid spermatogonia undergo several mitotic divisions forming primary spermatocyte, which is also diploid.

b) Primary spermatocytes then undergo meiosis in maturation zone and produces haplcid spermatids.

c) Zone of spermeogenesis where in the spermatids develops into the flagellated sperms. It is also called the zone of basal transformation. When cysts rupture, the sperms are released which then enter into the vas efferens and vas deferens and are then stored in the seminal vesicles. The sperms are held together in bundles by a gelatinous material when they pass from the ducts. These bandles of sperms are called spermatoderms. Insect sperms are usually with a narrow head and filamentous tail, both having same diameter the contractions of the vas deferens and ejaculatory duct muscles aid in the movement of sperms within the male reproductive system.

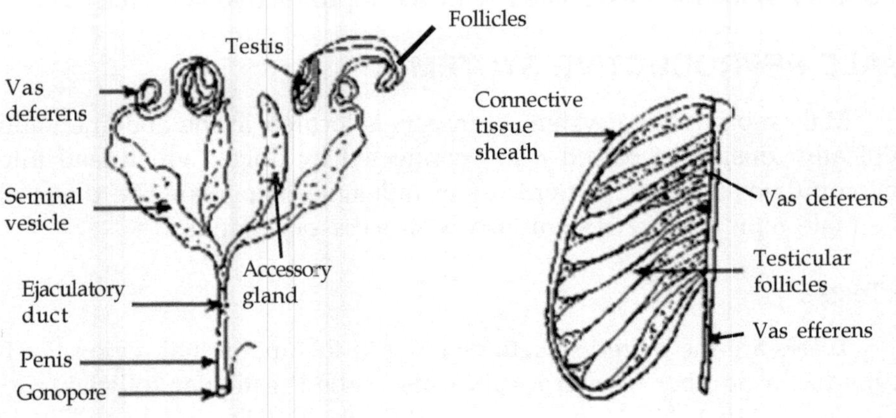

(A) Principle Male Reproductive Organ (B) Detailed Structure of Testis

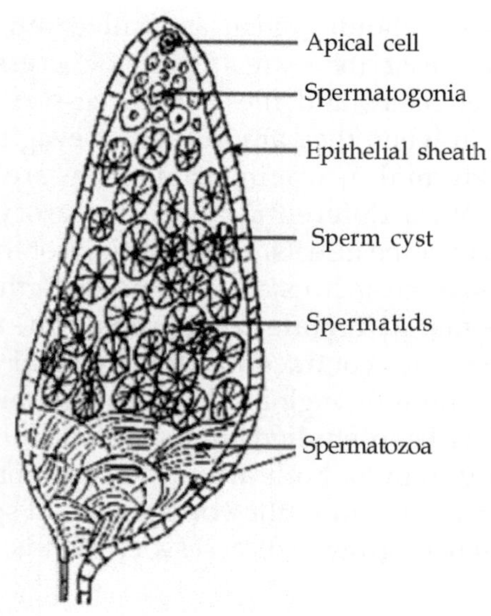

(C) Internal structure of Testicular Follicle

ii) Vas deferens

Vas deferens is a lateral duct of male reproductive systems, to which each testicular follicle communicates by a small duct called the vas efferens. The vasa deferentia from each testis unite at middle to form the ejaculatory duct. In some primitive insects, as many flies, each vas deferens opens separately to the exterior. Vas deferens is covered by a layer of muscles and connective tissue. In some insects, seminal vesicle is formed by the dilation of the distal end of the vas deferens; white sperms are stored for some time.

iii) Ejaculatory duct

Ejaculatory duct is formed by the union of two type vas deferens from each side. It ends in the penis or aedeagus at the gonopore. It helps in the propulsion of semen. It is lined with cuticle and has originated from the ectoderm of the embryonic layers. The walls of the ejaculatory duct are muscular and contractile. Ejaculatory ducts are very complex in those insects where spermatophores are formed *e.g.*, cockroach, locusts, etc. It may also have some glandular functions.

iv) Accessory glands

Many accessory glands are associated with the vasa deferentia or the seminal vesicles or with the ejaculatory ducts. In cockroach, there are large no. of accessory glands which are called mushroom glands or utricular glands. In *locusta,* there are 15 pairs of accessory glands. These are missing in silver fishes and flies. The accessory glands are either mesodermal in origin when they are formed as evaginations of the vasa deferentia, or ectodermal in origin, when they are formed from the ejaculatory ducts. Vasa deferentia and ejaculatory ducts have also glandular functions in some insects. The secretions of these glands contain chemicals that activate the sperms and also produce the spermatophores. It also makes the mated female for oviposition, accelarate oocyte maturation and stimulates contractions of the genital ducts which move the sperm, and formation of vaginal plugs which inhibit the subsequent in semination. It has been studied that seminal fluid production and growth of accessory glands, both are under hormonal control. When corpora allata was removed from the young males of some insect species, there was retardation in growth of accessory glands.

v) Spermatophores

In few insects, the sperms are passed in special gelatinous capsulls called spermatophores, instead of transferring them directly into the spermatheca of female. In spermatophores, the sperms are held together by the male accessory gland secretions. These are common in apterygotes, cockroach, grasshoppers, crickets, etc., and are rare or absent in Hymenoptera and some other neopterans. In the spermatopores, two or more layers are visible in the capsule having one or two sperm containing sacs. Spermatophores are formed before or during coupulation. Some moths produce spermatheca in female genitalia.

FEMALE REPRODUCTIVE SYSTEM

Female reproductive system is located in the abdominal region. It consists of paired ovaries connected to common duct or median duct by lateral oviducts. Genital chamber sometimes forms vagina, into which the common oviduct opens posteriorly. Bussa copulatrie develops from vagina, opens to the exterior and recieves male intromittant organ during copulation. Spermatheca are attached to the genital chamber of vagina, which store sperms and two accessory glands

i) Ovaries

Ovaries are mesodermal organs of egg production and are located bilaterally. Each ovary is composed of a number of ovarioles, which are its functional units. Ovarioles are travessed by a number of tracheas and is made of muscles and an epithelium layer. For the maturation of eggs in the developing follicles, oxygen is continuosly supplied by the trachea. The number of ovaries varies greatly in insects but it is constant for a particular species, from 1 in tse-tse-fly to over 2000 in queens of some termite species large species have greater ovarioles than small ones. A long terminal suspensory filament is formed by the union of long terminal threads arising from the distal end of each ovariole, in most of the insects. Sometimes, these filaments from two sides merge into a median filament, which attaches to dorsal diaphragm. The pedicle from the base of each ovariole forms a bulbous calyrs by uniting with the pedicle of other neighbouring overiols. A mature ovum or eggs, capable of being fertilized is formed by the process of oogenesis.

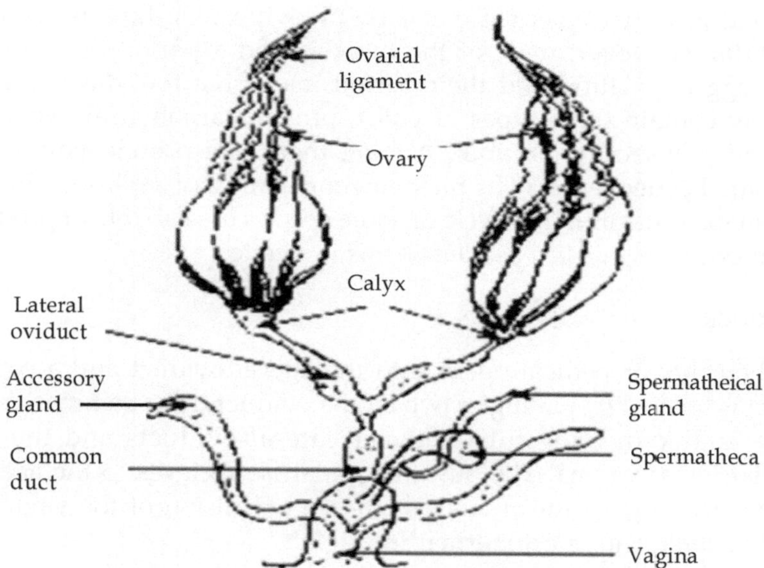

Female Reproductive System of General Insect

Type of Ovarioles

Based on the method of yolk deposition, ovarioles are of two types:

a) *Panoistic ovariole:* This ovariole has no trophocytes or nurse cells, and is the most primitive type of ovariole. Follicular epithelium surrounds each developing oocyte and follicular plugs are formed between adjacent oocytes. It is mostly found in apterygotes grasshoppers, termites, dragonflies, somebeetles etc.

b) *Meroistic ovariole:* It contains trophocytes. These ovarioles may be polytrophic or telotrophic, depending upon the site of trophocytes. In polytrophic ovarioles, each developing oocyte is in direct association with the nurse cells, *e.g.*, wasps, butterflies, moths, etc.

In telotrophic ovarioles, cytoplasmic strands or nutritive cords connect the nurse cells to various oocytes, nurse cells being present in the terminal germenial layers *e.g.*, heteropterans insects.

Vitellogenesis

Yolk deposition, which occurs into the oocytes in the lower parts of the ovariole, results in a very rapid increase in size of oocyte. *E.g.*, In Drosophilla, the volume of oocyte increases about 1, 00,000 times during vitellogenesis. Ovulation rate is determined by the rate of vitellogenesis. In both the insects, which do not feed as adults, yolk deposition in oocytes occurs during late larval or pupal stages? In such a situation, the eggs are deposited just after emergence and no preoviposition time is needed by the adults. However, most of the insects need a period of maturation before egg deposition and their period may from few days to weeks. Egg may contain three types of yolks, protein-carbohydrate yolk, lipid yolk and polysaccharide yolk. Among them, the protein yolk is most abundant. Paedeogenesis The phenomenon in most of the insects in which the immature stages in life cycle of some species are capable of producing mature oocyte *e.g.*, *Micromalthus-debilis* (a beetle).

ii) Oviducts

The calyx or pedical opens into the lateral oviduct and a common oviduct is formed by joining of two lateral oviducts. The common oviduct serves as a connecting tube between lateral oviducts and the bursa coupulatrise or vagina. The bursa coupulatrise recieves to the aedeagus during intromission and it is a pouch-like expansion of the vagina it is lined bycuticle and is ectodermal in origin.

iii) Spermatheca

It is a bag-like structure, associated with vagina, in which sperms are stored before fertilization. Its number varies in different insects. It is joined by a spermathecal gland or itsepithelium itself becomes glandular, and secrete spermathecal fluid that nourishes the sperms.

iv) Accessory Gland

Accessory glands usually open into the apical portion of bursa coupulatrix, there are usually one or two pairs, and vary in structure and function. An adhesive cementing material, secreted by these glands, holds the eggs together to the substratum in ootheca. The aquatic insects have accessory glands which secrete a gelatinuous secretion around the eggs.

The accessory gland secretion of some insects provide nutrition for the incubating larvae *e.g.*, tse tse fly. This is a defensive secretion in some insects to paralyse the prey *e.g*, honey bees. Insemination Semen, during copulations, is deposited in bursa coupulatrix or common oviduct. In insects, fertilization is internal, and its takes place generally by coupulation. Female insects are mostly polyandrous *i.e.*, mates several times in her life. However, some others mate only.once in life-time *i.e.*, they are monoandrous *e.g.*, bees, wasps, flies etc. Similary, males of most of the species are polygynous. When semen is deposited in the vagina of female, either it is free or is accompanied by one or more spermatophores from male to the female. In some species, semen is deposited in lateral oviducts or even directly into the spermatheca. Some insects, *e.g.*, bugs, haemocoelic insemination occurs, in which sperms are transferred by male into the female haemocoel, by perforating its vaginal wall with a spine at the tip of the aedeagus. Many of the semen and sperms are then phagocyt by female and some of them reach the varies. However, in apterygota, the male first deposits the spermatophore on to a suitable substrate and then female comes and inserts it in her vagina by herself.

Fertilization

Fertilization in insects occurs in three ways:

i) Release of sperms from spermtheca.

ii) Entry of sperm into the egg.

iii) Formation and fusion of the male and female pronuclei.

Sperms stored in spermatheca usually survive for several months or years in monoandrous insects, and at desired time, are used to fertilize the eggs in the median oviduct, bursa copulatrise or vagina. While as in poly androus species, the sperm storage in spermatheca is for time being. How the sperms release from the spermatheca, is not well understood. In most insects, sphinctor muscles surround the opening of spermathecal duct, these muscles guard the opening of the spermathecal ducts, and these muscles are intrun regulated by neuro secretions from neuro secretory cells embedded in these muscles. Many female hymenoptera reproduce arrhenotoleously, *i.e.*, males develop by parthenogenesis while females develop by sexual reproduction. The decision of the female, regulating the sperm release, is influenced by many factors. They are:

i) Temperature

ii) Host-complex

iii) Host size

iv) Parental car

v) Host distribution etc.

After ovulation, egg is rotated in the reproductive tract in such a way, that the micropyler region is in close proximity to the sperm release site. The sperms then respond chemotactically and move towards micropylar region of egg. In most of the insects, more than one sperm enter the egg membrane but only one fuse with its nucleouas, forming the zygote by the fusion of the male and female pronuclei.

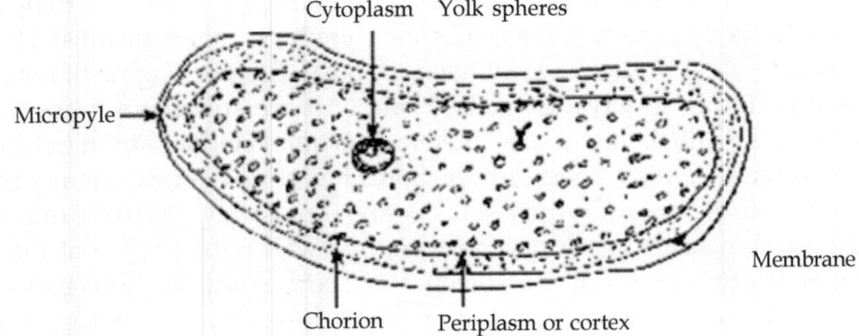

Section of a Typical Insect Egg, Structure

Female Reproductive Organs of palaemone

It consists of a pair of ovaries and a pair of oviducts.

i) *Ovaries:* Two, white, compact and sickel shaped ovaries are placed above the hepato pencrease and below the pericardial sinus and heart. Immature ova lie towards centre, while mature ova lie towards periphery of ovary.

ii) *Oviducts:* These are short, wide and thin walled tubes and both the oviducts arise from ovary at the middle of its outer border. It runs vertically downwards and opens by a female genital aperture.

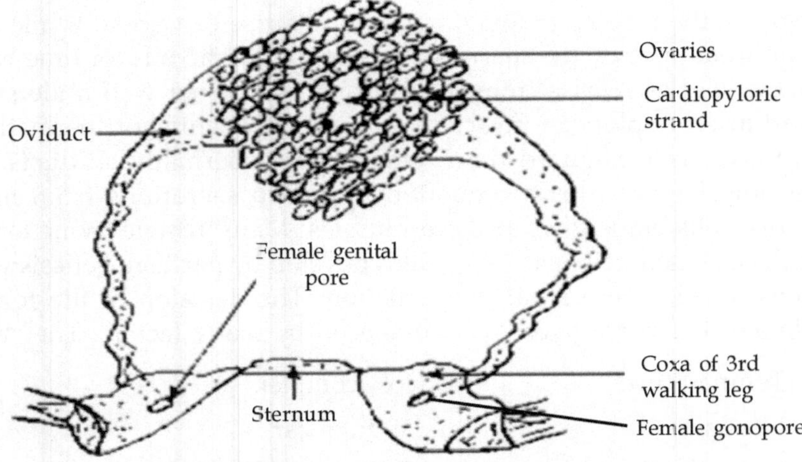

Female reproductive system of palaemone (Prawn)

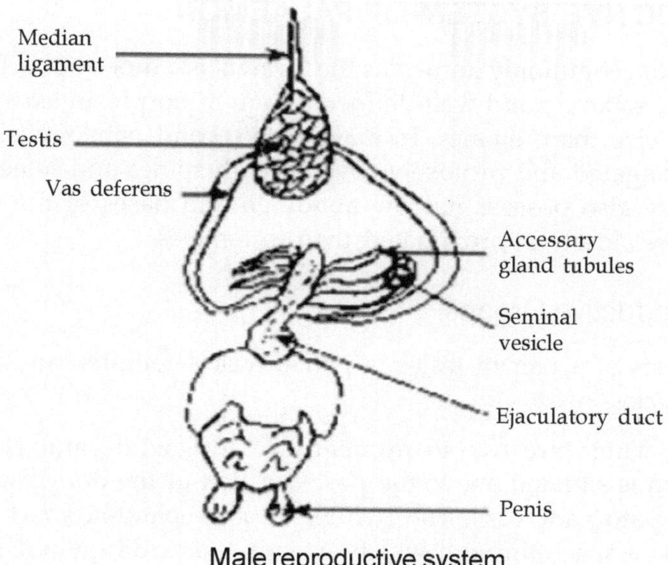

Male reproductive system

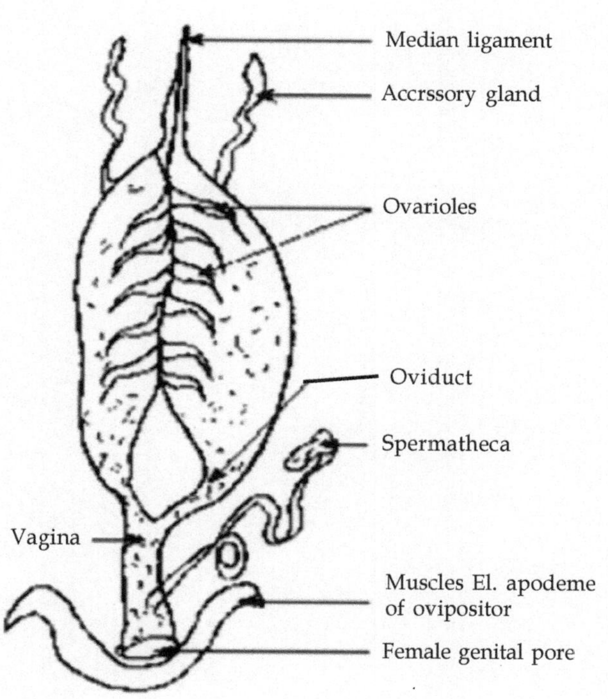

Female reproductive system

Grasshopper or locust

REPRODUCTIVE SYSTEM OF PALAEMON

Palaemon, commonly known as the 'prawn' is a dioecious arthropod, *i.e.,* sexes are separate and well-defined sexual dimorphism exists. Males are large in size than females. In males, the second pair of chelate legs are more elongated and profusely covered with spines and setae then in females, they also possess narrow abdomen and bases of the thoracic legs are more closely approximated than in females.

Male Reproductive Organs

It consists of a pair of testes, a pair of vas deferentia and a pair of seminal vesicles.

i) *Testes:* There are two in number, are soft, white and elongated structures situated above the posterior half of the dorsal surface of the hepato pancrease. The two testes are separated apart so as to enclose a space through which extends the cardio-pyloric strand.

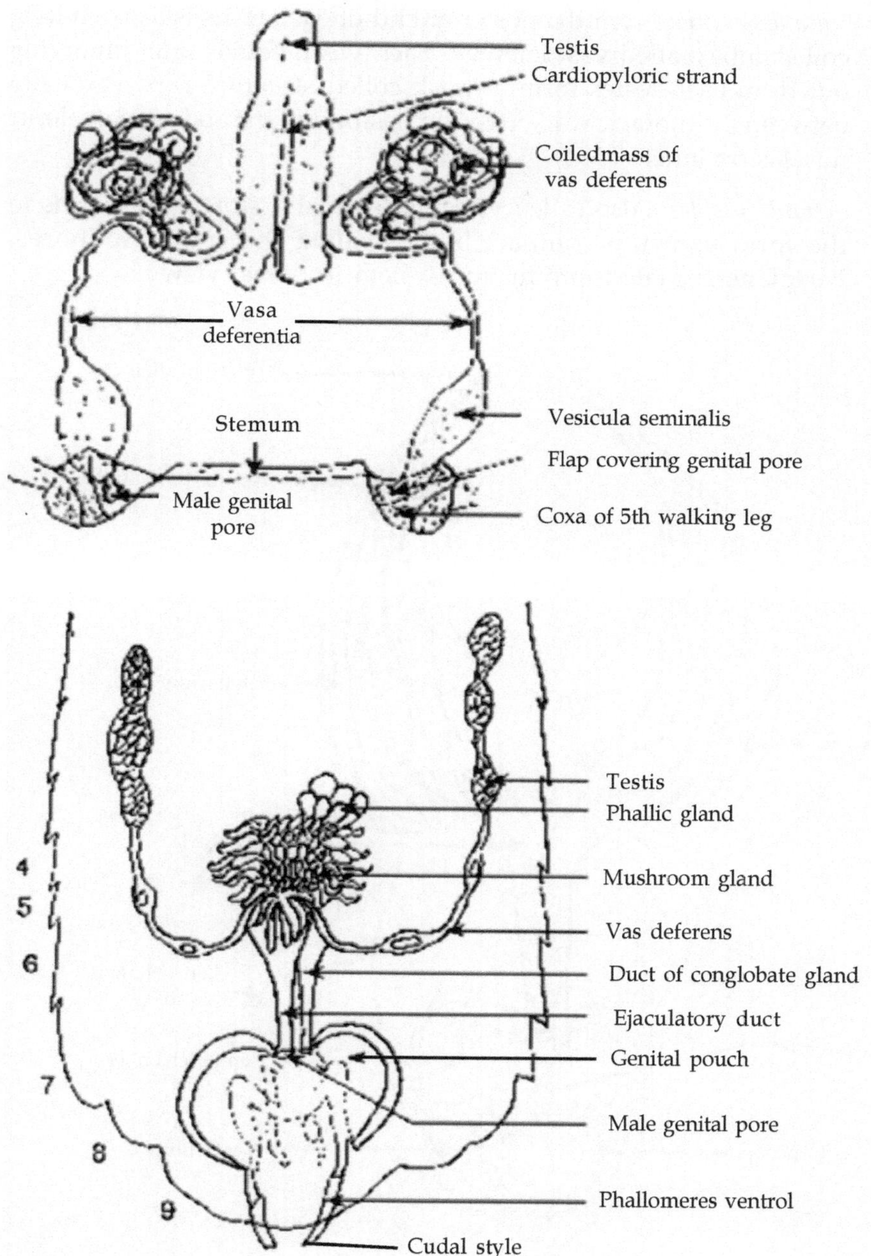

Male Reproductive Organs = Periplaneta: Male reproductive system in dorsal view structure

ii) *Vasa deferentia:* From the posterior end of each testis is a size a long coiled tube, called vas deferens. Each vas deferens after emerging out from each testes form a much coiled structure runs vertically between the thoracic wall on the outer side and the abdominal felerur nuselas on inner side.

iii) *Seminal vesides:* Also called vesicula seminalis, stores the sperms in the form of white compact bodies called the spermatophores. **Periplaneta:** Male reproduction system in dorsal view.

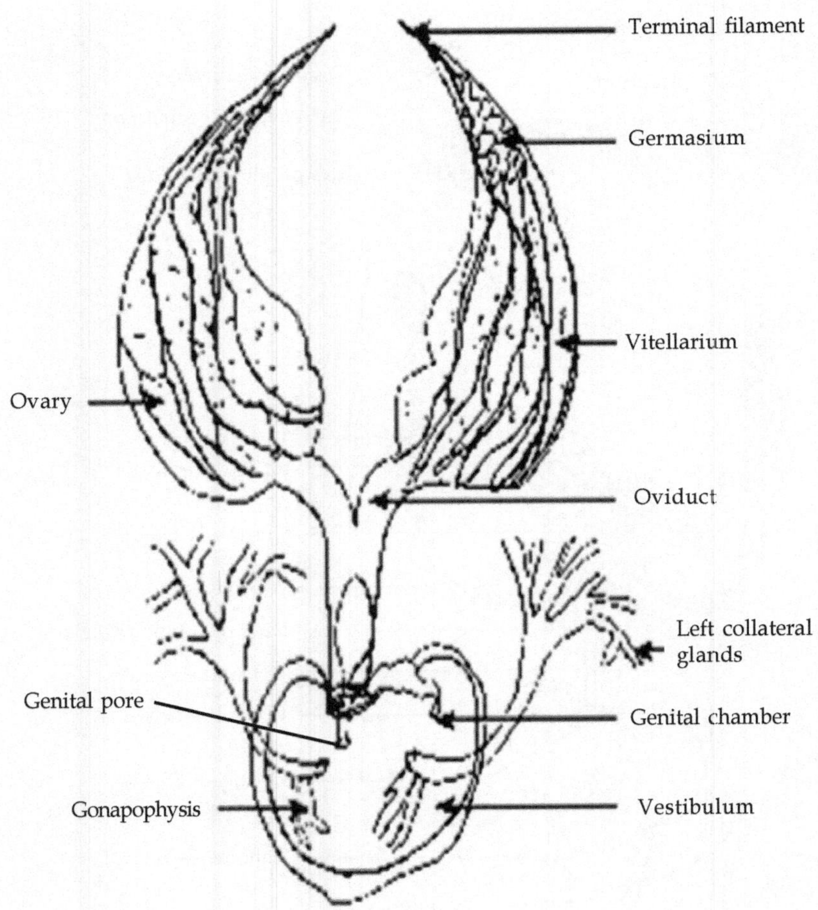

Periplaneta: Female Reproductive System

8

Respiratory System of Grasshoppers

The respiratory system or tracheal system is involved in gaseous exchange in the insect with the environment. In this system there is a system of internal tubes, the tracheae that directly transport the oxygen to the parts of the body. Therefore, it does not use the circulatory system as the vehicle for gaseous exchange. The trachae open outside through segmental pores called spiracles having some system of closing and opening. Except proturans and some collembolans all insects possess tracheal system for respiration. These insects live in moist habitats where gaseous exchange takes place directly with the environment via the integument.

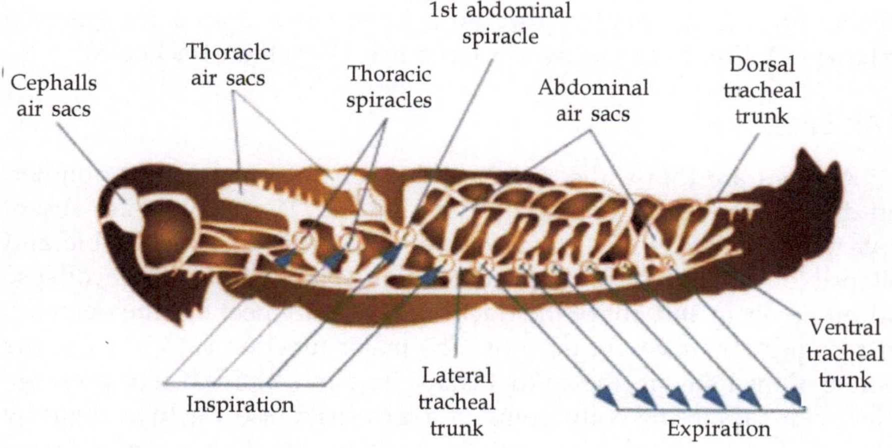

Gaseous Exchange in Grasshopper

i) Tracheae

The tracheae, ectodermal in origin, are tubes that communicate with the outside by spiracles. From each developing trachea, branches are given off to the various organs including the wings. The tracheae are circular or somewhat elliptical in cross section. Histologically the tracheae are similar to the integument, being composed of a layer of epithelial cells that secrete a cuticular layer, the intima. Chitin is absent in the smaller tracheal branches. The intima is thrown into a series of usually spiral folds around the lumen called taenidia. The taenidia provide strength to the tracheae and protect it against collapse with changes in pressure. The intima is shed along with the old integument during each moult. Although tracheae are resistant to compression in a transverse direction, they can he stretched longitudinally to some extent. It helps the insects in which the abdomen becomes greatly distended with food, *e.g.,* blood-sucking insects.

ii) Tracheoles

The tracheoles are intracellular smallest branches of the tracheal system, ranging in size from 0.2m to 1.0m in diameter, and are the place where gaseous exchange takes place. The very fine taenidia (10-20 mm) are retained during moulting unlike tracheae. A trachea typically ends with a tracheal end cell, the tracheoblast, which gives rise to several tracheoles that are all a part of this cell. The tracheoles are very intimately associated with the tissues or organs that have a high metabolic rate and high oxygen demand. These tissues or organs include the flight muscles, ovaries, fat body, gut epithelium, Malpighian tubules, and rectal papillae collectively the tracheoles provide a huge surface area for gaseous exchange. A fifth instar silkworm larva has 1.5 million tracheoles.

iii) Air Sacs

Air sacs are thin-walled tracheal dilations of varying size numher, and distribution found mainly in flying insects. The taenidia are absent or very poorly developed, therefore, air sacs are quite distensible and collapsible. Whenever the demand of O_2 increases, the air sacs collapse and pump air in and out of the tracheal system to meet out the demand. Air sacs perform several functions. The major function is to increase the volume of the tidal air (the an that is inspired and expired). The presence of large air sacs in the body cavity of a terrestrial insect help in flight by reducing the specific gravity and of aquatic insects it gives some degree of buoyancy. Air sacs also provide space for growth of internal organs. It helps in heat conservation in large insects that generate high temperatures for flight. It improves the haemolymph circulation tn the flight muscles. Air sacs also form the tympanic cavity of the hearing organs of various insects (*i.e.,* tymbal of the cicada).

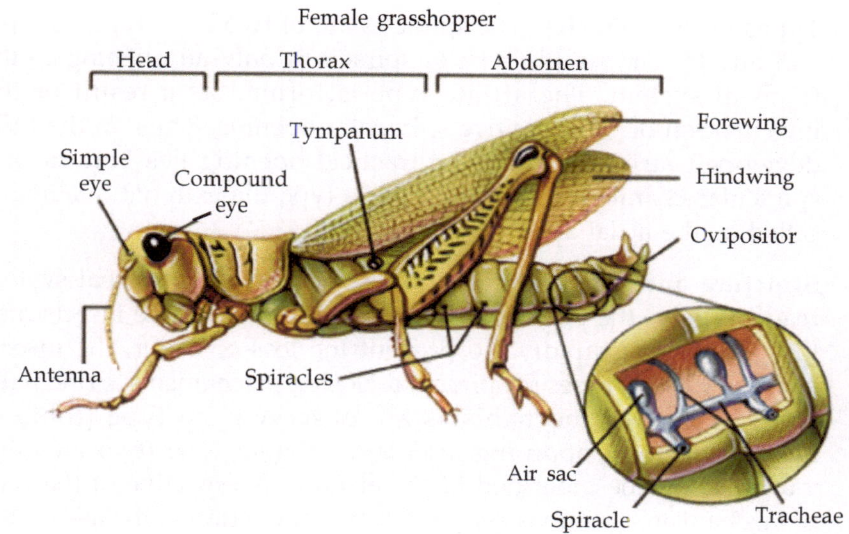

The tracheal system of insect

iv) Spiracles

Spiracles are the external openings of tracheae they are lateral in position; usually on the pleura usually a pair of spiracles are present in a body segment.

1. **Number and distribution of spiracles:** With the exceptions of some dipluran insects, the maximum number of spiracles found in insects is ten pairs, two thoracic and 8 abdominal. On the basis of the number and distribution of spiracles, the respiratory system is classified as:

   a) *Polypneustic:* It's the least 8 pairs of functional spiracles of insect. The Holopneustic-10 pairs of spiracles in the insects is (1 mesothoracic. 1 metathoracic and 8 abdominal), *e.g.*, cockroaches. *Peripneustic*-9 pairs of spiracles (1 mesothoracic, 8 abdominal), *e.g.*, some fly larvae. *Hemipneustic*-8 pairs of spiracles (1 mesothoracic, 7 abdominal), *e.g.*, some fly larvae.

   b) *Oligopneustic:* 1 or 2 pairs of functional spiracles *Amphipneustic* 2 pairs of spiracles (1 mesothoracic 1 post abdominal), *e.g.*, a larva of moth flies; *Metapneustic*-1 pair of functional spiracles (post abdominal), *e.g.*, mosquito larvae: *Propneustic*-1 pair of spiracles (mesothoracic). *e.g.*, Dipteran pupae.

   c) *Apneustic:* No functional spiracles, *e.g.*, chironomid larvae.

2. **Types of the spiracles:** The spiracles are of two basic types: simple and atriate. The simple type ol spiracle is only an opening to the tracheal system. The atriate type is formed as a result of the invagination of the primitive spiracular opening. Thus, in the fully developed atriate spiracle the tracheal opening lies bottom of a spiracular chamber, or atrium. In this type the external opening is called as the atrial aperture or orifice.

3. **Structure and closing/opening of spiracles:** The tracheal system readily allows the passage of water and due to this the insects may lose water very rapidly. To prevent the loss of water, the insects evolved various types of spiracular closing mechanisms. Principally types of closing mechanisms are observed, lip type (folds of integument form opposing lips) and valvular type (two movable valves lies at the inner end of the atrium). Irrespective of the type of mechanism, closure is carried out by contraction of the associated muscles. Most often the atrial wall is lined with tiny hairs, which form a felt chamber that filter out dust. In flies, beetles and moths the spiracle is covered by a sieve plate having large numbers of fine pores that not only prevents the entry of dust but also the water (in aquatic insects) in the tracheal system. The spiracle is also associated with certain glandular tissue that secretes lubricants for the movable parts of the spiracle closure mechanism the lubricants may prevent water from entering the tracheae, and may improve the seal of the closure mechanism. Longitudinal section is the spiracle of a louse showing the dust catching spines.

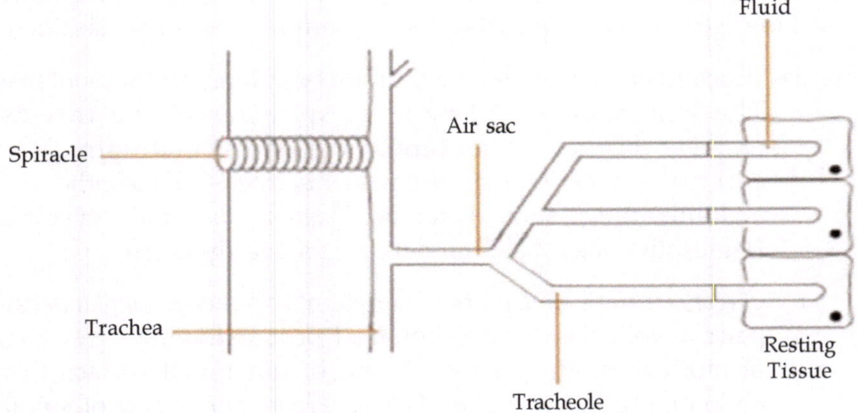

Tracheal System of grasshopper

v) Types of the Tracheal System

Except most collembolans, many proturans, and certain endoparasitic wasp larvae other insects possess tracheae. Tracheae along with spiracles, air sacs, and tracheoles compose the respiratory or ventilatory system. The organisation of tracheae may be comparatively simple, as in some springtails in which tracheae arise from each spiracle, but do not connect to any other tracheae. Although the basic pattern of tracheation is genetically determined, new tracheae and tracheoles can be induced to develop if an insect is reared in an atmosphere with a very low oxygen content. New tracheae and tracheoles do not develop between successive moults, but on demand changes in the distribution can occur at the time of moulting. Major tracheal branching patterns are species-specific and are often very similar among members of a given family ot order Based on the presence or absence and functional or non- functional nature of spiracles, there are principally two types of tracheal systems, open and closed, with a variety of modifications within each type.

1. **The open tracheal system:** Most of the insects have open tracheal system which is characterised by the presence of one or more pairs of functional spiracles.
2. **The closed tracheal system:** Many aquatic and endoparasitic insect larvae do not possess functional spiracles and the gaseous exchange occurs directly through the integument, *e.g.*, Chironomus *larva* is a mayfly nymph.

vi) Mechanism of Gaseous Exchange and Ventilation

Two types of gaseous exchange are observed in insects: diffusion or passive ventilation and active ventilation.

1. **Diffusion or passive ventilation:** Simple diffusion from the outside of smaller insects and from well-ventilated air sacs in larger insects can supply sufficient oxygen to the body tissues to maintain life. It is a passive form of ventilation in which the gases are not pumped in the tracheae and tracheoles. Diffusion is also regulated by the opening and closing of the spiracles. The spiracles respond to decreased O_2 or increased CO_2 in the air by remaining open for longer periods of time. Spiracular opening and closing are under both neural and hormonal control.
2. **Active ventilation:** In larger and active insects passive ventilation does not bring adequate amount of oxygen to the tissues In these insects, air sacs, if present, and larger tracheae are often ventilated by rhythmical pumping movements of the body which is called as

active ventilation Peristaltic waves over the abdomen telescoping or dorsoventral flattening of the abdomen, and, in some, movements in the thorax or even protraction and retraction of the head cause the inspiration and expiration of gases. In addition, heartbeat and the gut movement, assist ventilation by pressing against adjacent tracheae. Both tracheae that are oval in cross section and air sacs are collapsible and hence can serve to increase the volume of tidal air.

3. **Elimination of CO_2:** In tissues CO_2 diffuses about 35 timers more rapidly than O_2. Because of this, CO_2 is much more likely to be eliminated from the body through the tracheal linings and integument than is O_2 to be absorbed along the same routes. Thus, although most of the CO_2 produced by respiration is eliminated via the tracheae and tracheoles, some of it may escape through the general body surface of soft-bodied insects and the intersegmental membranes of hard-bodied insects. In some insects CO_2 is not continuously eliminated through the spiracles, but in regular bursts, while O_2 consumption remains constant. Between these bursts the spiracles remain fully closed or half-closed. The spiracles open completely during a CO_2 burst. As oxygen is removed from the tracheoles and tracheae by respiration, at least a portion of the carbon dioxide produced presumably goes into solution as bicarbonate in the haemolymph. A CO_2 burst probably indicates the previous buildup of CO_2 (in the haemolymph and tracheae) to a threshold above which complete spiracular opening occurs. The ability to release CO_2 periodically allows an insect to keep its spiracles partially or entirely closed most of the time and hence is thought to be an adaptation that favours the conservation of water by diminishing the rate of transpiration.

vii) Respiration in Aquatic Insects

Many insects spend all or part of their lives in an aquatic environment. These insects must either be able to utilise O_2 in solution or have some means of tapping a source of undissolved O_2 whether it be at an air-water interface or from aquatic vegetation.

1. **Use of dissolved O_2 in water:** Aquatic insects with closed tracheal systems depend entirely upon the diffusion of dissolved O_2 through the integument (cutaneous respiration) These insects obtain O_2 in a variety of ways. Mayfly and damselfly nymph possess tracheal gills, which are integumental evaginations covered by a very thin cuticle and are well supplied with tracheae and tracheoles. Such gills are usually abdominal. Other aquatic insects with closed tracheal

systems possess spiracular gills (*e.g.*, pupae of some dipterans), or cuticular gills.

2. **Use of aerial O_2:** Aquatic insects having open tracheal systems obtain O_2 at the surface of water and for this they come at the surface of the water periodically. Some aquatic insects may remain submerged for an indefinite period of time and have certain, tructures that help them in obtaining aerial O_2.
   a) *Respiratory siphon:* The larvae of mosquitoes and *Eristalis* possess posterior spiracles on siphon that penetrates the water surface and get atmospheric O_2. In Eristalis the siphon is telescopic and may extend to a length of 6 cm or more.
   b) *Hydrofuge structures:* Hydrofuge structures are usually made up of hairs and are resistant to wetting by water, *e.g.* in *Notonecta.* Thus, when an insect approaches the surface, the cohesive properties of water cause it to be drawn away from the hydrofuge areas. These structures are generally associated with particular spiracles. Certain dipterous larvae have peristigmatic glands that secrete fatty substances in dipterous larvae have peristigmatic glands that secrete fatty substances in the immediate neighbourhood of the spiracle and make it hydrofuge. Hydrofuge structure also severs to keep water out of the tracheae when the insect is submerged.
   c) *Air stores:* Many aquatic bugs and adult beetles carry air stoics in the form of bubbles or films into which spiracles open these air stores are often held in place by a pile of erect hydrofuge hairs, but in some insects the body is so shaped that it forms a storage area without the use of hydrofuge hairs. Films or bubbles of air would obviously be a temporary source of O_2 if the insects were forced to remain submerged. To replenish air stores water scorpions (Hemiptera, Nepidae) use a caudal siphon, a long hollow tube extending from the rear of the body.
   d) *Physical gills:* Many aquatic insects that carry stores of air are able to replenish the O_2 without surfacing. This is performed by the air store acting as a physical gill. As the O_2 in reserve is used up, a point is reached where the partial pressure of oxygen is less in the air store than it is in the surrounding watet. At this point oxygen diffuses from the water into the air store. Nitrogen in the air store does not readily diffuse into the water and hence tends to keep the air store from collapsing. Air stores usually make an insect positively buoyant and may play a role in hydrostatic balance.

e) *Plastron:* Insects that are able 10 remain submerged lot a longer period usually possess a structure known as a plastron. A plastron is a very thin layer of gas held firmly in place by tiny hydrofuge hairs or other very fine cuticular networks. The latter are typically associated with spiracular gills. Hair plastrons are found on adults of certain aquatic beetles (*e.g., Elmis*), nymphs and adults of the aquatic bug *Aphelocheirus,* and adult females of the wingless moth *Acentropus.* Plastrons composed of cuticular networks are found in the larval and/or pupal stages of certain beetles and flies. Unlike typical air stores, the gas layer held by a plastron cannot be displaced by water. The spiracles open into the plastron, and it functions in a manner similar to a physical gill except that it does not require repletion by a visit to the surface.

f) *Use of plant surface:* Insects obtain O_2 from submerged vegetation in a variety of ways. Many are able to hold bubbles on the surface of plants by means of hydrofuge structures. Others penetrate the tissues of submerged plants by biting into them or by inserting a specialised ventilatory structure into the intercellular air spaces, *e.g.*, certain mosquito larvae other flies, and some beetle larvae.

viii) Respiration in Endoparasitic Insects

Most of the endoparasitic insects are parasitic only in the immature stages, *e.g.*, parasitic wasps. The environment of these insects presents problems similar to those of the aquatic insects. In several insects, the tracheal system is non-functional and respiration is cutaneous, gaseous exchange occurring directly between the tissues of the parasite and body fluids of the host. Some endoparasitic insects have tracheal gills. The larvae are a Cotesia (Hymenoptera. Braconidae) possess caudal vesicle which is an everted structure of the hindgut. The wall of the vesicle is very thin and is associated with the heart so that O_2 passing in is quickly earned round the body. Others depend, at least partly, on aerial O_2 obtaining it either by tubes or other structures that communicates with the tracheal system and that extend out of the host to the atmosphere.

ix) Respiratory pigments

The haemolymph does not generally function as an oxygen carrier. However the haemolymph of certain chironomid larvae (bloodworms), the endoparasitic botfly larva (*Gasterophilus*), and the backswimmers (*Anisops*) contain haemoglobin as a respiratory pigment. Under conditions of high oxygen tension this haemoglobin is saturated with oxygen and thus does not serve as a carrier. However, under conditions of low oxygen tension the haemoglobin is unsaturated and hence available to carry oxygen.

x) Central Nervous Control of Respiration

The ventilation involves the muscular system and the nervous system for activation of and coordination over the muscular movements. As ventilation is an oscillating system, it is regulated by specific ganglia that contain the essential neural net-work and mechanisms for generating coordinated rhythmic outputs of the motoneurons that stimulate the muscles expressing desired behaviour. These motor output programmes are usually instinctive, stereotypic, and species specific. It is believed that each ganglion, which registers some input to the muscles that produce either ventilatory movements or spiracular closure, has its own local control center, but one of these may serve as the pacemaker. The ventilatory pacemaker in *Schistocerca gregaria* appears to be in the metathoracic ganglion, whereas in dragonfly's nymph it resides in the last abdominal ganglion.

9

Introductory Nervous System of Insect

In insects nervous system is highly developed and serves to coordinate the activities of its various systems. Units of this system are known as neurons. Which carry information in the form of electrical impulses from external and internal sensilla (Sensory cells) to appropriate effectors *e.g.*, glands, muscles and glail cells. Nervous system of insects may be divided into three parts.

1. Central nervous system
2. Sympathetic or visceral nervous system
3. Peripheral nervous system.

1. Central Nervous System

CNS constitutes the main division of nervous system and the basic unit of the central nervous system are brain and a double chain of neutral nerve and having segmental ganglia joined by lateral and longitudnal connectives. Central nervous system is divisible into the following:

A. The brain or cerebral ganglion

B. The suboesphangeal ganglion

C. The ventral nerve chord

A. **Brain:** Brain is very complex and is located just above the oesophagous between the supporting apodemes of tentorium. It is formed of optic, antennary and inters calary ganglia in the embryo. It contains three distinct lobes from dorsal to ventral.

   i) Fore brain or protocerebrum

   ii) Mid-brain or Dentocerebrum

   iii) Hind brain or Tritocerebrum

- Protocerebrum represents the fused pair of optic ganglia and is the largest and most complex part of brain having distinct cell masses and regions of neuropile: optic lobes, ocellar centers, central body, Protocerebral Bridge, pons intercerebralis and corpora-pedunculata. It contains optic nerve and compound eyes. The mid brain is formed by the fusion of antennary segment which innervates the antennae. Hind brain is formed by the ganglia of the third of interealary segment of the head. It connects the brain to the stomatogastric nervous system via the frontal ganglion and to the ventral chain of ganglia via the circum oesophageal connectives. Connecting both sensory and motor elements.

B. **Suboesophageal ganglion:** It is the ventral ganglionic centre of head and is formed by fusion of ganglia of madibular, maxillary and labial segments.

C. **Ventral nerve cord:** It consists of a series of ganglia lying on the floor of thorax and abdomen. The ganglia of adjoining segments are joined by paired conectives. The first three are thorax ganglia and rest are located on abdomen. Each ganglion gives principle nerves.

2. The visceral nervous system

It is also called sympathetic nervous system. It consists of the frontal ganglion located in front of the brain and connected to tritocerebrum by a pair of fibres. Anteriorly it gives off a frontal nerve which passes to the clypeus posteriorly passing through oesophagous to the brain and expands into a hypocerebral ganglion.

3. Peripheral nervous system

All the nerves radiating from the ganglia of the central nervous system are sympathetic nervous system. Nerves contain motor fibres. Just beneath the hypodermal layer there are bipolar and multipolar nerve-cells whose prolongation forms a network of nerves. The nerves are named on the basis of particular organ where they go *e.g.*, eyes and ocelli names, manidibulae, maxillary and labial nerves, leg nerves and wing nerves etc.

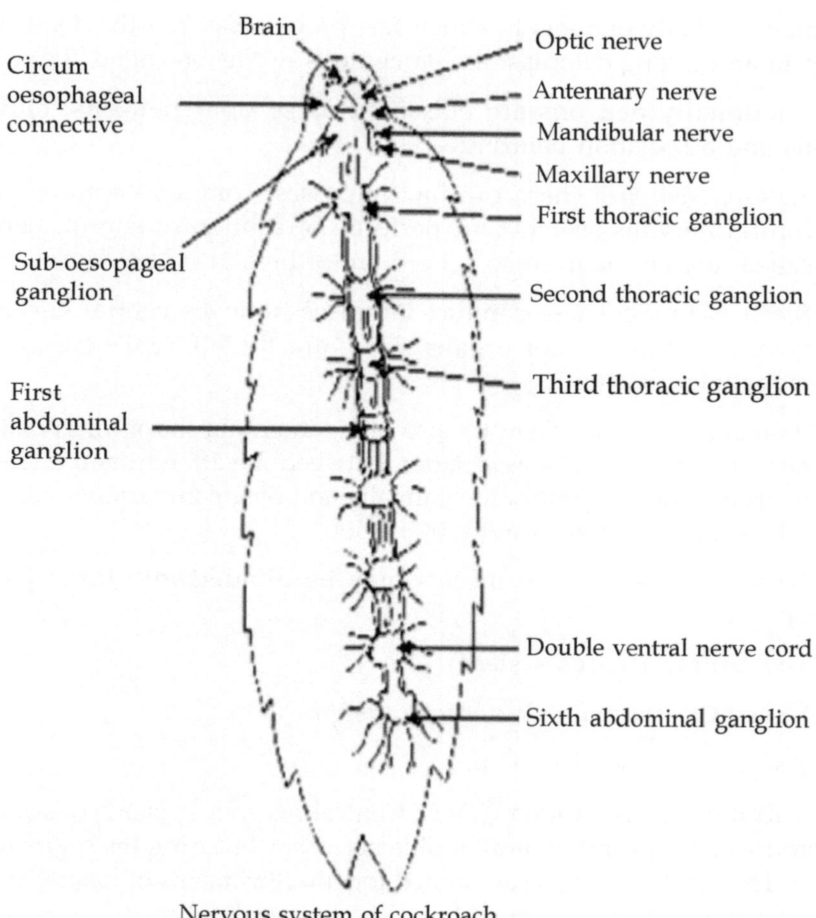

Nervous system of cockroach

Nervous System of Insects

REFERENCE TO GRASSHOPPER

Evolution of multicellular in animals made it imperative to develope some system for the control and coordination of the activities of numerous cells in the body. This requires collecting information about changes in the external environment and transmitting this information to the internal cells located at a distance from such cells. Information has also to be received about changes in the external environment and then transmitted to the cells not immediately near the site of change. Like other metazoa, the Nervous system of insects forms the main site of control and coordination. It is well developed in insects and serves as the connecting link between the sense organs and the effect of the organs. Neurons form the Morphological, physiological functional unit of nervous system. Neurons are ectodermal in origin and typically consist of a prominant nucleated cell body or soma to which fine processes called the dendrites belong to and a long cytoplasmic filamentous extension called Axon.

Functionally neurons are classified as sensory neurons, motor neurons and association neurons:

i) *Sensory Neurons:* These conduct impulses from a receptor to the central nervous system and perform perception of stimuli hence called afferent neurons too. They are of bipolar nature.

ii) *Motor Neurons:* These conduct impulses from the central nervous system to the effector organs. They are the efferent neurons of monopolar nature.

iii) *Association Neurons:* They are also called interneurons or interpolated neurons as they make association between the afferent and efferent neurons. They lie within the gangalia and bear comparatively small cell bodies occupied heavily by nuclei.

The nervous system of insects may be divided into three parts (*Locust*):

i) The central nervous system.

ii) Visceral or sympathetic nervous system.

iii) Peripheral nervous system.

Central Nervous System (CNS): Central nervous system consists of a anterodorsal brain and ventral double nerve cord bearing the segmental ganglia. The series of ganglia are joined together by means of longitudinal and transverse strands of nerve fibres. The longitudinal cords are termed as connectives and they serve to join a pair of ganglia with these which precede and succeed it. The transverve fibres or commissures unite the two ganglia of a pair. It includes a brain or cerebral ganglion, suboesophageal ganglion and the ventral nerve cord.

a) *Brain:* The insect brain represents the supraoesophageal ganglion and it lies above the oesophagus in the oculo, frontal region of the head. Its bilobed structure and a median furrow separate the two lobes. The insect brain is divided into three dorsoventral regions – the protocerebrum or fore brain, the deutocerebrum or mid brain and the *tritocerebrum* or hind brain. Its (Brain) externally covered with the air sacs and fat body.

The insect brain is covered with brain sheath called *cerebral neurolemma*. Its composed of two layers the outer neural lamella non-cellular one and the inner perineurium – cell layer. *Cortex* forms the peripheral region of the brain and other ganglia. It's filled with monopolar, motor neurons and the glial cells. *Medulla* is composed of a mass of axons of varying diameters, their terminal arborizations investing glia and tracheales. Its compactly filled with the neuropile. The protocerebral neurophile contains some distinct neurophile masses *viz.*

i) Pons cerebralis or Protocerebral Bridge. It is a medial neuropolar body. It is situated in the frontal region of the protocerebrum as a littice neuropile behind the pars intercerebralis - transversally.

ii) Corpora pendaculara or Mushroom bodies, these are two mushroom shaped bodies, each on either side of the pars intercerebralis.

iii) Corpus centrale or central body its a median mass of neuropile at the center of the protocerebrum.

iv) Corpora Ventralia or Accessory lobes, A pair of corpora ventralia lie in the ventrolateral region of the protocerebrum and both the ventral bodies are connected with each other by a protocerebral transverse commisure.

v) Corpora optica or the optic lobes. They are the optic lobes and donot represent the ganglia. Each lobe contains a group of three neuropile masses outerlamina ganglionaris, middle-medulla externa and *inner*–medulla interna.

In *Locust* or Grasshopper Brain lies dorsally above oesophagus and consists of 3 pairs of fused supra oesophageal ganglia. Brain is distinguished into:

i) Protocerebrum

ii) Deutocerebrum

iii) Tritocerebrum

The protocerebrum represents the fused pair of optic ganglia. It forms the greater part of the brain and innervates the compound eyes and ocelli. The deutocerebrum is formed by the fusion of antennary segment which innervates the antennae. The tricocerebrum is formed by the ganglia of the 3rd of the intercalary segment of the head. It is divided into two small widely separated lobes which are attached to the dorsal lobe of the deutocerebrum and receive nerve fibres from the latter. The tritocerebrum lobes are joined together by means of postoerphageal commissure which passes immediately behind the oesophagous. Brain chiefly serves as a relay centre which receives stimuli from sense organs and thus directs the movements of body. It also exerts an inhibiting influence but does not seem to coordinate muscular activity for the animal to jump, walk and fly even after removal of the head or brain.

b) *Circum-Oesophageal Connectives:* Brain is connected by a pair of circum-oesophageal connectives around oesophagus to a ventral sub-oesophageal ganglion.

c) *Sub-oesophageal ganglion:* It also represents fusion of three pairs of ganglia (Mandibular maxillary and labial) and innervates the mouth parts. It also plays some role in maintaining balance. Both the brain and Sub-oesophageal ganglia contain neursecretory cells that secrete a hormone which activates other hormones to control moulting and some other aspects of matamorphosis.

d) *Ventral nerve cord:* A double ventral nerve cord arises from the hind end of sub- oesophageal ganglion and extends down the mid ventral region of the body. It bears bilobed ganglia, three thoracic and five abdominal. Third or metathoracic ganglion is the largest as the first three abdominal ganglia are also fused with it. Similarly, 5th abdominal ganglion is the largest representing last four pairs of abdominal ganglia fused together. All segmental ganglia are coordinated by nerve fibres running in the ventral cord. Besides, each ganglion independently controls the movements of its respective segment or segments including appendages. Thus, a several nerve cord does not result in paralysis as in vertebrates. So that a severed thorax can walk off by itself and an isolated abdominal segement can performbreathing movements.

Peripheral Nervous System Some bipolar and multipolar neurons form a complex network beneath the integument above the segmental muscles and upon the surface of the gut. The distal processes divide enormously over the innervated structure and their centripetal processes are terminated into the ventral ganglia. Peripheral nervous system comprises nerves arising from ganglia of central nervous system. From protocerebrum arises a pair of stout optic nerves and three fine ocellary nerves, which supply the compound eyes and ocelli respectively. Deutocerebrum sends off a pair of antennary nerves to antennae. From tritocerebrum arises a pair of labro- frontal nerves, each of which divides into a labral branch to labrum and a frontal ganglion, connective to frontal ganglion. From suboesophagal ganglion arises a pair each of hypopharygeal, mandibular maxillary and labial nerves and a few pairs of cervical nerves to neck.

From each ganglion of ventral nerve card arises several paired nerves and supply their respective segments. Metathorax and first 3 abdominal segments are supplied by nerves from metathorax ganglion, while last four abdominal segments receive nerves from the last (fifth) abdominal ganglion.

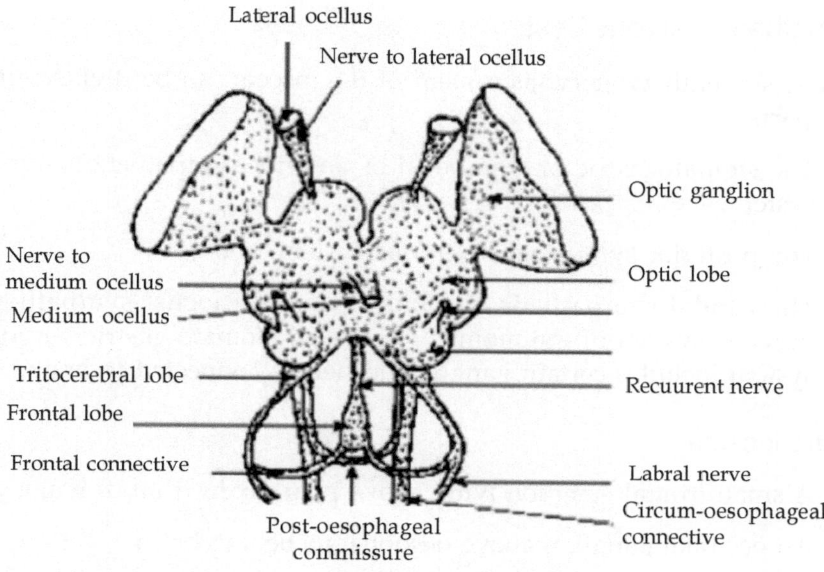

Dorsal view of Brain

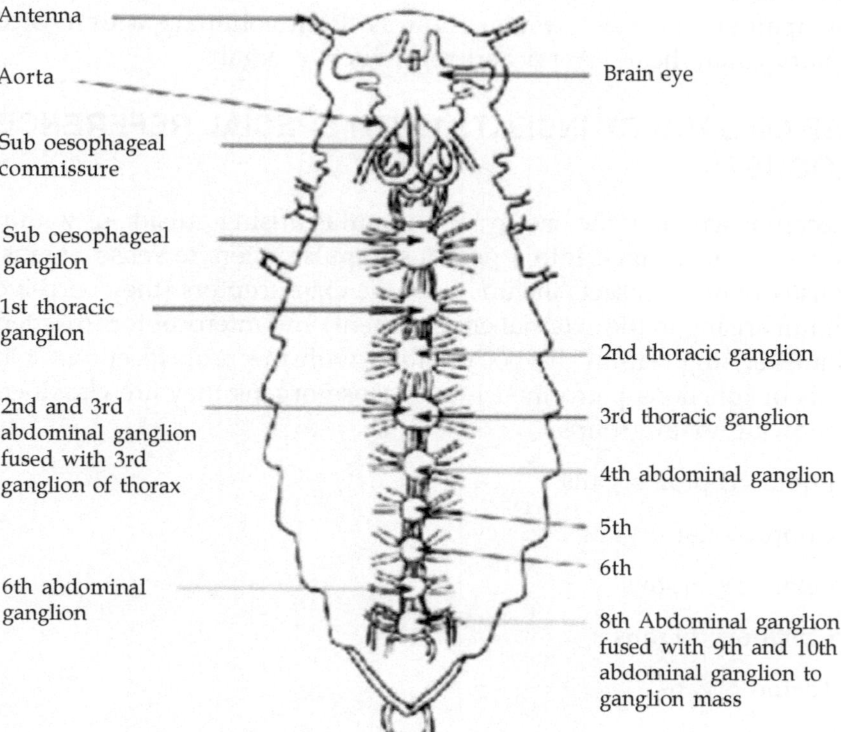

Insect Nervous System of (Locust)

Sympathetic Nervous System

The sympathetic nervous system of the insects can be divided into three parts:

a) The stomatogastric or stomodeal or anterior sympathetic nervous system,

b) The posterior sympathetic nervous system, and

c) The caudal sympathetic nervous system. In locust sympathetic nervous system or autonomic, visceral or stomato gastric nervous system includes certain ganglia and nerves connected to brain.

Ganglia include

i) A small frontal ganglion lying above pharynx in front of brain.

ii) An occipital ganglion above oesophagus behing brain.

iii) A pair of ingluvial ganglia in front of gastric caeca.

Sympathetic nervous system controls the involuntary activities of alimentary canal, heart, Aorta and reproductive organs.

SENSE ORGANS OF INSECTS (WITH SPECIAL REFERENCE TO LOCUST)

Receptor whereby the energy of a stimulus arising outside or within the insect is transformed into a nervous impulse refers to sense organs. The sense organs of insects include both the exteroceptors (they perceive the stimuli arising in the external environment) and interoceptors (excited by stimuli arising within the body) along with the proprioceptors. On the basis of functions performed by the sense organs they are classified into following main groups:

1. Photoreceptor organs
2. Chordotonal organs
3. Olfactory organs
4. Gustatory organs
5. Tactile organs
6. Humidity, temperature and receptors of insects.

1. Photoreceptors

The most primitive photoreceptive organs of insects are represented by the paired small papillae on apex of the head of the muscoid maggots. In current age insects following photo receptors are worthy to note.

A) **Ocellus:** These refer (ocelli) to the simple eyes of insects. In insects two types of ocelli are found.

i) *Dorsal Ocelli:* They are found in the nymphs of hemimetabolous insects and in almost all adult insects. Typically, they are three in number and situated in a triangle. They are shifted to the vertex in Diptera and hymenoptera. In *locust* the median ocellus is located on the frons while the paired ocelli on the vertex. A typical insect ocellus bears following main parts:

i) *The cornea:* Is a transparent cuticular structure which is arched or raised to form the outer investment of the ocellus?

ii) *The corneagen layer:* This layer secretes and supports lens and is composed of colourless transparent cells.

iii) *Retina:* It's composed of visual cells which are sensory neurons. An ocellus consists of a group of photoreceptor cells or retinules each ending in nerve fibre which leads to the brain. Outer end of each photoreceptor cells forms a rhabdome cuticle covering the group of photoreceptor cells, forms a thick, biconvex transparent lens.

iv) *Pigment cells:* Between the retinulae ocelli contain accessory cells bearing pigments, have been noticed in few cases

Functions: The ocelli are simple light sensitive organs activating the central nervous system and thus play a key role in maintaining the diurnal rhythm. They are well adapted for quick perception of light, changing in its intensity. They are however, mostly incapable of forming images because of wide angular separation of their rhabdoms and the images are focused far behind the retina.

ii) *The stemata or lateral eyes:* As the name itself indicates that these eyes are lateral in position. They differ from dorsal ocelli as they are innervated from the optic lobes of the brain. They vary in number from a single pair to the six pairs amongs tholometabolous immature forms. The adult insects generally have a single pair of lateral eyes on the head.

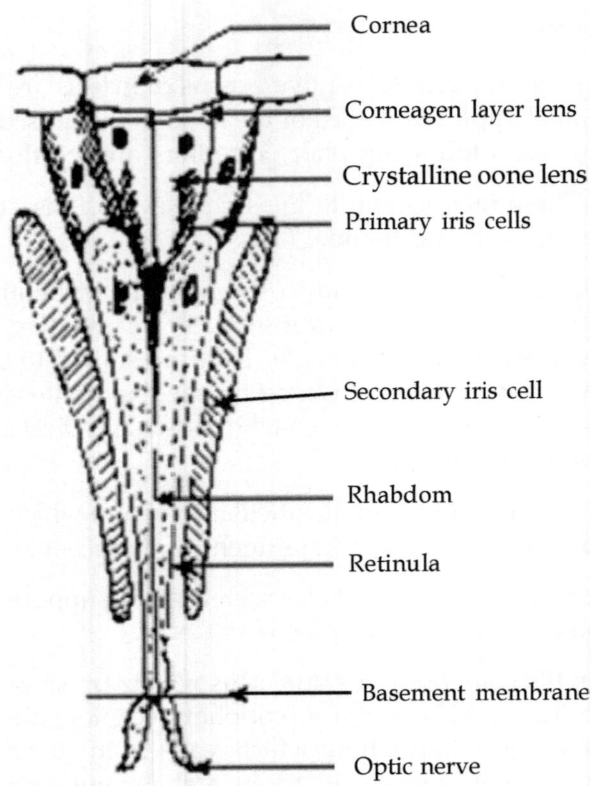

Longitudinal Section of Ocellus in insects

Functions: They are capable of differentiating the shapes and colours of the objects and can form images on the rhabdoms.

B) Compound eyes: The compound eyes of insects are composed of large number of structural units called ommatidia. The compound eyes are chief visual source or organs of Machilidae and majority of the pterygotes. The cornea of compound eyes is divided into a large number of facets often equal to that of ommatidia. It should be noted that the facets size is mostly proportional to the square root of the height of the eye.

The basic difference between the ocellus and the compound eye is that:

i) The ocellus bears a dome shaped common lens for all visual cells while in the compound eye each visual cell is provided with an independent lens.

ii) Lens is cuticular.

iii) Pigment cells are lacting.

iv) Retina is made up of only four retinula cells.

v) Axons form ocellar nerve.

vi) Ocellar nerve terminates into the par intercerebralis.

vii) Cone is absent.

viii) Ocelli donot form image of an object.

Fundamentally each ommatidium consists of a dioptric or optical apparatus receiving the light and a sensory or retina forming the image. In addition to it various components of ommatidium include.

i) *Cornea:* It's the outermost part of the ommatidium. It is transparent, colourless and biconvex modified cuticular area often termed as a facet or lens. It is shed during each moult.

ii) *Corneagen layer:* They are the epidermal or hypodermal cells lying behind the cornea. A group of two corneagen cells secrete a single lens.

iii) *Cone cells:* Just Beneath the corneagen layer of cornea, there are four distinct cells which secrete a transparent body known as crystalline cone or semper cells.

iv) *Retinula cells:* The crystalline cone is followed by a long retina forming the basal part of an ommatidium. It is formed from a group of similar seven pigmented visual cells. The visual cells secrete an internal optic rod or rhabdom. The portion of rhabdom formed by each cell is termed as rhabdom. These rhabdomeres of all seven retinula cells extend the whole length of the Retina. In addition to these cells there are deeply pigmented cells which surround the cells of crystalline cone and the corneagen layer, refered as primary iris cells. Enlongated pigment cells which surround the primary iris cells and the retinula are termed as secondary iris cells.

2. *Chordotonal organs:* These are also called auditory organs which are usually compound structures made up of a spindle shaped group of scolopophores. They are intravisceral mechano receptors. Both the ends of these organs are attached to the integument of insect. Each scolopophore consists of (*i*) a cap provided with a terminal ligament apically and acts as a scolopale cap, (*ii*) the scolopale is formed within an envelope cells that contains sensory cuticle rod, (*iii*) the sensory

cells as the third cells of the scolophore which extends dendrite process apically. It runs through the vacoule, scolopale and finally terminate into the end knob of scolopale. These organs are generally found in the abdominal appendages and wing bases of insects. It is reported that besides sound receptors they are also used for various purposes including proprioception and the reception of internal pressure changes and mechanical vibrations.

The highly specialised chorodontonal organs are:

i) **Johnston's organ:** The Except collembola and Diplura all other insects posses the Johnston organ in the second antennal segment. This organ was first seen and reported by Christopher Johnston which consists of a variable number of radially arranged sensilla. These are attached at one end to the second and third intersegmental membrane and at the other end to the wall of the second segment. The axons of the sense cells of these organs run back and enter the antennal nerve. The antennae also contain large number of companiform and chordotonal sensilla. The Johnston organs respond to the movements of the antennal flagellum and thus act as not merely the photoreceptors but also perceive air currents during flight, vibrations through substratum during walking and through

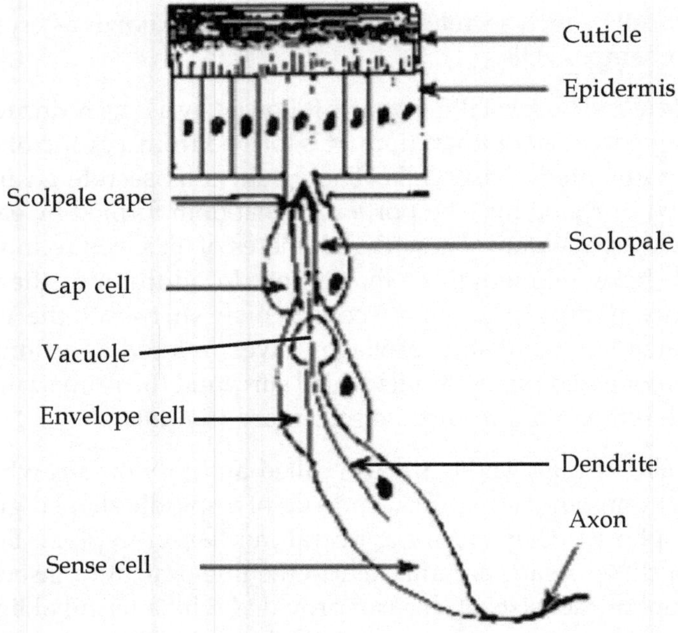

Chordotonal Organ: Basic Structure

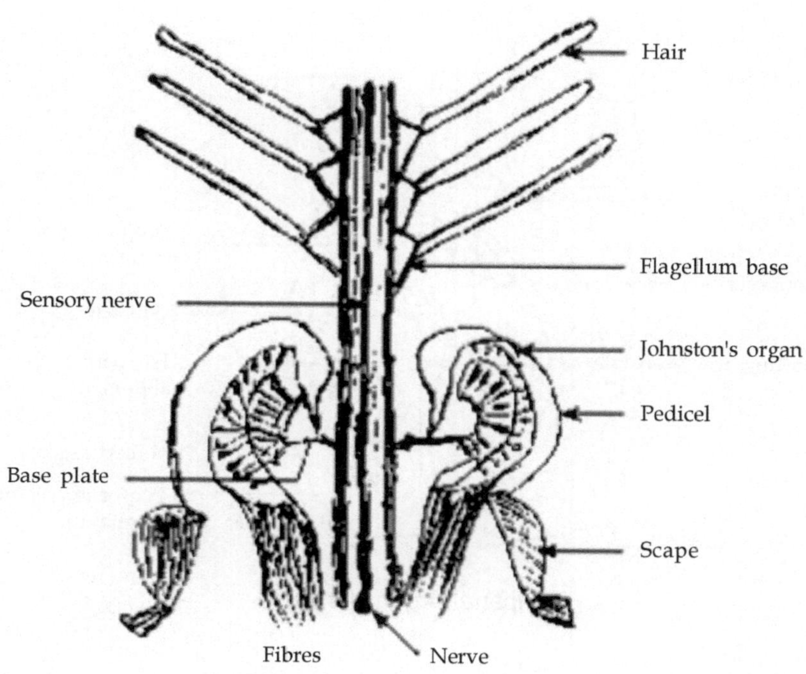

Johnston's organ of male mosquito (after Atrium 1963)

water surface at swimming.

ii) **Tympanal organs:** These are paired structures made of thin cuticular membrane known as the tympanum. They are associated with tracheal air sacs and chordotonal sensilla. These tympanal organs have reported to occur at the base of each fore tibia in the members of tettigonoidea and grylloidea. In grasshopper the tympanal organ is situated on each side of the first abdominal tergum and externally surrounded by a cuticular ring. Internally there is a sensory organ known as mullers organs which is formed by a group of numerous

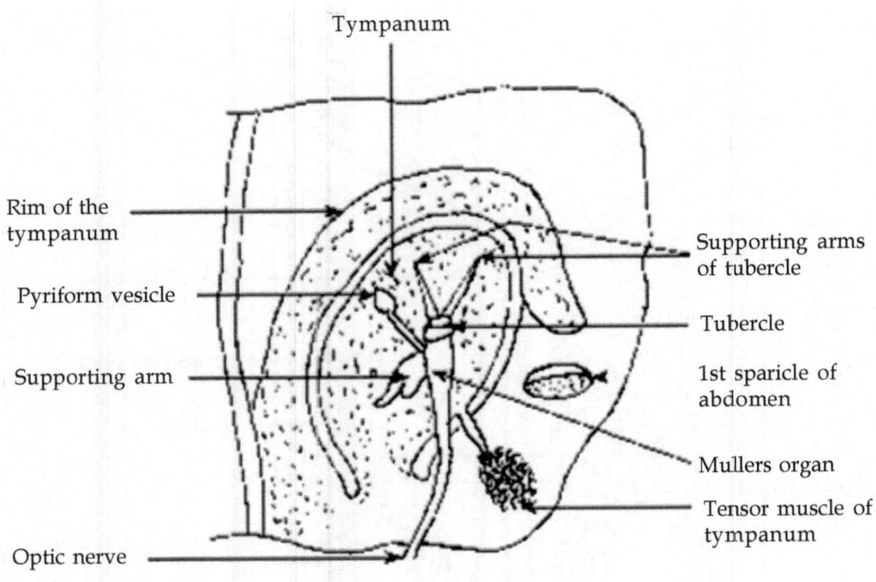

Tympanal Organ of Insects

scolopophores and connected to the metathorax ganglion by auditory nerve

iii) **Olfactory organs:** These sense organs provide sense of smell that is mediated by chemical stimuli in a gaseous state at low concentration. There are, however some receptors which appear to mediate responses to chemicals in both gaseous and aqueous phases in some aquatic beetles. Antennae are supplied with the organs of smell – *Locust.*

iv) **Gustatory organs:** These provide sense of taste that is mediated by chemical stimuli acting with liquids at high concentration. In the terrestial insects, the gustatory chemoreceptors are mostly confined to the mouth parts, wall of true oral or pharyngeal cavity rarely on the tarsi and distal end of tibia as found in the butterfly moths, honey bee etc.

v) **Tactile organs:** Tactile hair sensitive to touch are located on various body parts, particularly the antennae, palps and distal l*e.g.* segments. They are simple articulated sensory hairs and are commonly called sensilla. They are made up of cuticle and articulated within the socket with the body wall. The trichoid sensillum is formed by two cells, the hair by the trichogen cell and the socket by the tormogen cell. Both of these cells are epidermal in nature. The tactile hairs of the antennae and lower segments of legs perceive earth-born vibrations in terrestrial insects and water surface vibrations in aquatic insects.

vi) **Humidity and temperature receptors**: These are the most important organs specially meant for controlling the desications in insects. Humidity receptors have been identified by pendiculus in the form of a peculiar tuft like cuticular sensilla and in many species as basiconic, trichoid and placoid sensilla. With the help of these sensilla ants predict about the rains well in advance that is why they migrate to safer places along their eggs, larvae and pupae prior to onset of monsoon.

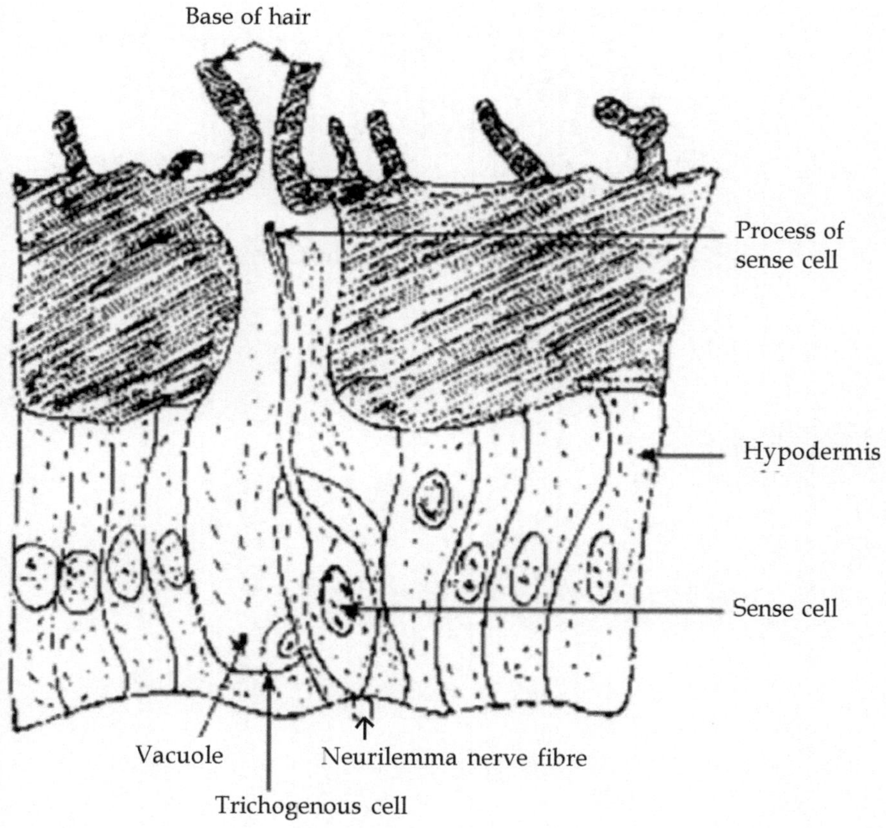

Tactile Hair of Cereus

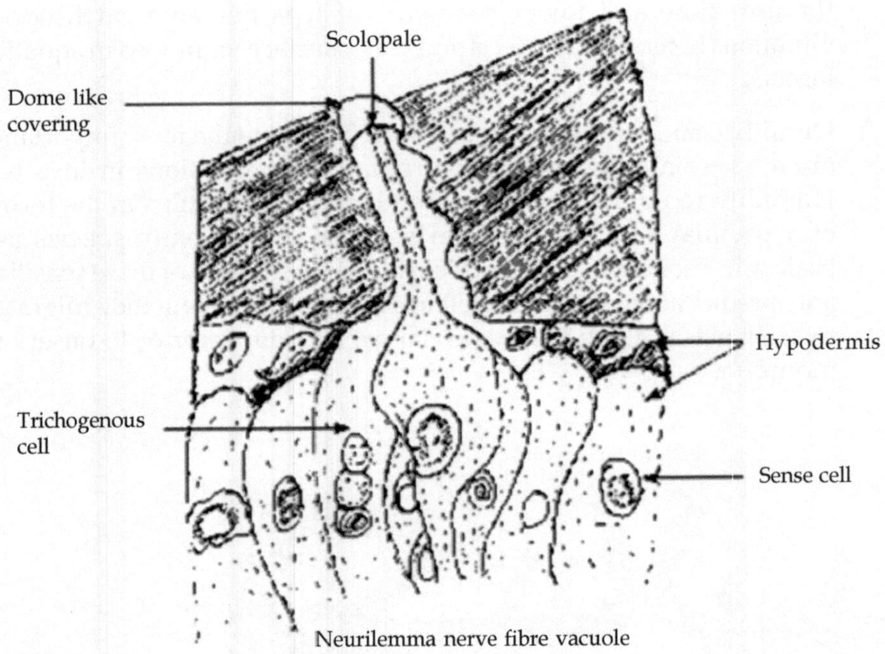

Companiform Sensillum from the Cereus

10

Exocrine and Endocrine Glands of Insect

Gland: They may be single cells or small aggregates of secretory cells or can be complex glands. Gland is a bunch of cells that secret wide variety of substances with a wide variety of functions. All the cells present in a gland are secretory to some extent secreation is a main function of glands.

Exocrine Glands Types: Exocrine glands are generally of ectodermal origin and are widely scattered over the insect. Exocrine glands release their secreations through ducts into the lumen of various internal organs. Example: Accessory reproductive glands, salivary gland, silk gland. Gland is mainly composed of secretory epithelial cells. These are simple unicellular, bicellular, multicellular, complex multicellular (compound glands). These glands are ectodermal in origin and are secreated over the insect body.

Examples. Wax glands, lac glands, cephalic glands, poison glands.

i) Glands pouring (secreting structural materials). Some insects secrete substances that are used in building their houses (beehive) as protecting them (lac or cocoon).

Following glands are recognized to secrete such substances

1. **Wax glands:** There glands are mostly found in homoptera and honey bees. These are uni or multicellular. Structures presents in integuments wax is secreating either in a form of powder or filament. The wax secreted by dermal glands located ventrally between the overlapping sternal plates of the fourth and the seventh abdominal segment is used to build a honey comb. The wax is secreated as fluids and through fine cuticular pores an accumulating and hardening in the form of thin plates which are removed by the hind legs.

2. **Lac gland:** Lac is secreted by certain coccoidea like tachardiidae and in particular by Kerria Lacca which produce commercial lac. Lac is produced in huge quantity by female and in fewer amounts by male as a protective covering.

3. **Silk glands:** Silk is secreated by labial glands in the form by *fibrogen* provide larval shelters in cocoons.

   Example - Lepidoptera and Trichoptera order.

ii) Glands secreting defense materials. Insect's secret a variety of chemical substances that are used to defend them from natural enemies. These substances have a pungent smell and other repellent properties. Some of these glands discussed below:

1. **Repugnatarial glands:** The repugnatarial glands secrete toxic substances which are variable in number location and morphology. The secretions of these glands initiate the escape response among the predators and other potential enemies. The chemical are secreted from the storage reservoir associated with glands.

   Example - Bombardier beetles eject a hot spray containing quinones from glands in the posterior part of explosion. Formic acid is secreted by ants which are used as defence substance.

2. **Poison glands:** These are peculiar to bees and wasps were they are modified accessory reproductive glands associated with the ovipositor or sting. Secretions are generally mixture of several substances like melittin and Kinin. Lepidopteran larva are provided with epidermal poison glands associated with seta or spines which when broken, discharge secretions in man causing utricaria.

3. **Pygidial glands:** In certain beetles, these type of glands are located at the posterior end of the body and open near the anus. It secretes pungent or corrosive materials.

4. **Frontal glands:** It is a sac-like organ, which typically open through the frontal pore. It represents a unique defense specialization characteristic only for advanced termite families of rhinotermitidae, termitidae and serriter mitidae. Its role has been well described only in soldiers. It manages the blosynthesis, storage and discharge of defensive and alarm substances.

5. **Alarm pheromone glands:** Aphids secrete triglycerides and other acids that are liberated through the cornides. When the aphids are attacked by predators and acting as alarm pheromone this causes the aphids in the variety vicinity to drop of the plants.

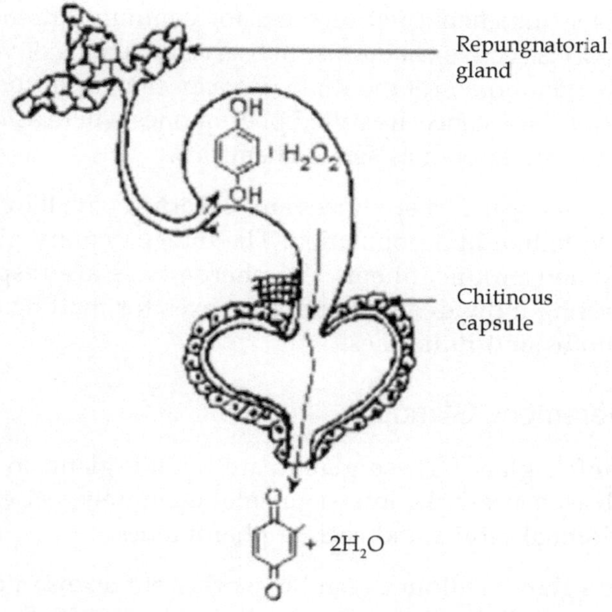

Repugnatorial gland of bombardier beetle

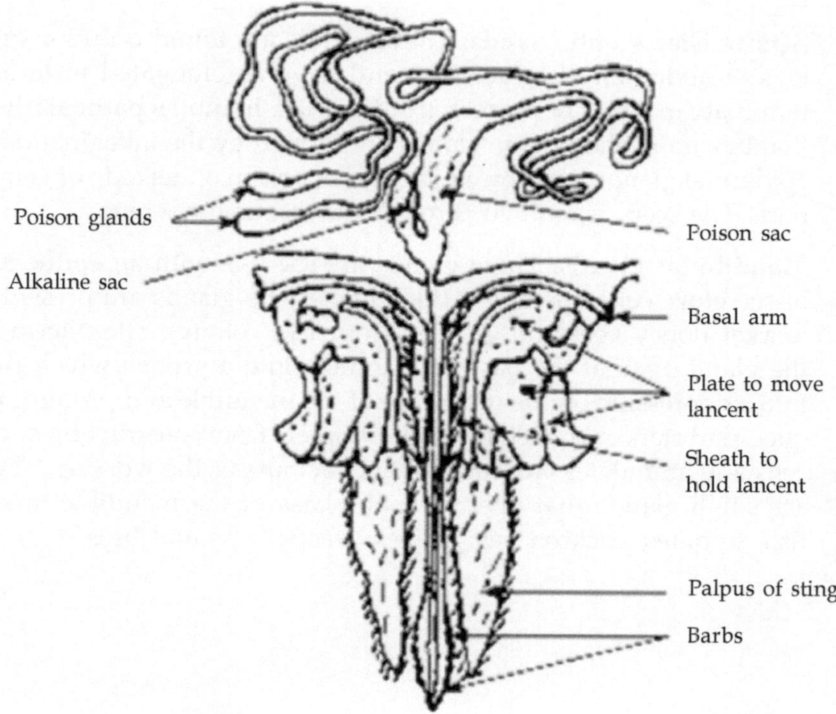

Poison gland associated with sting apparatus of honeybees

iii) Chemicals (infochemicals) secreted for communication by glands. Many insects secrete chemicals that serve as signals of various sorts for other membranes of the same (intraspecific) or different species. Intraspecific substance are called pheromones where as interspecific substances are known as semio chemicals.

a) *Pheromone gland:* Pheromones are conserned with the coordination of individual in a population. The integumentary glands of the abdomen produce them. The pheromones are responsible for attracting individuals of opposite sexes for matting particularly in moths and Butterflies.

Types of Pheromone Glands

1. **Nassanoff's gland:** These glands are mainly gland in the worker honey bees beneath the intersegmental membrane between 6th and 7th abdomnal tergits and secrete pheromones.
2. **Dufour's gland:** Dufour's gland in worker ant opens into the poison gland near the base of sting. It is small simple sac with their glandular walls and a delicate muscular sheath. The gland secretes trail pheromone.
3. **Aphrodisiac scents glands:** These glands are found on the wings, legs or abdomen of butter flies and have an elongated form and terminate in a row of processes or fimbriae. In moths particularly in Bombyx mori pheromone glands are formed by the invagination of epidermal glands lined by cuticle and opens on either side of genital pore. The scent is emitted through the terminal opening.
4. **Mandibular glands:** These glands are sac like with an epithelium of secretory cells lined by a thin cuticle. The glands are present in worker honey bees and Queen (Mother of colony). The duct from the gland open at the base of mandible into a groove which runs into a depression on the inner face of the mendible and produce the queen substance which inhibits the workers from constructing queen cells and stimulates various other behaviours of the workers. These are small glands that open near the base of the mendible in star fish, termites, cockroaches, beetles, wasps, ants and bees.

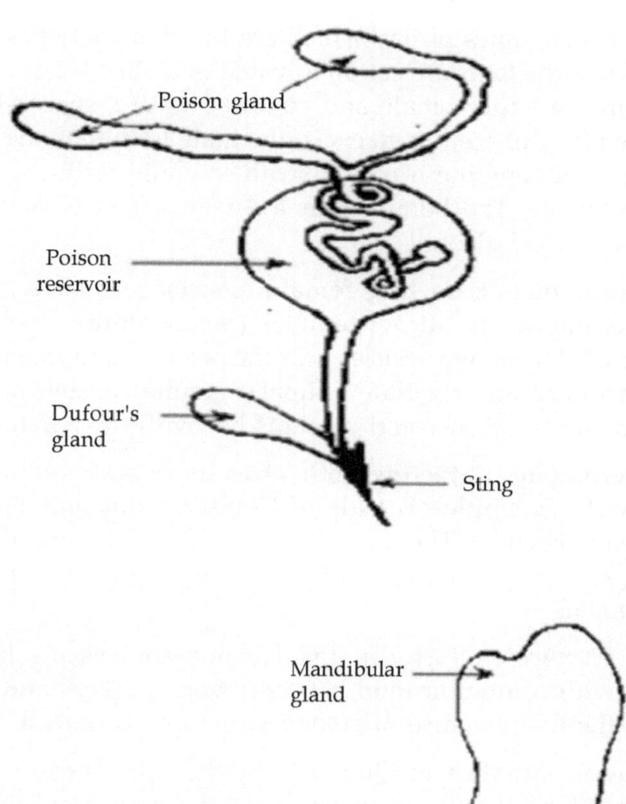

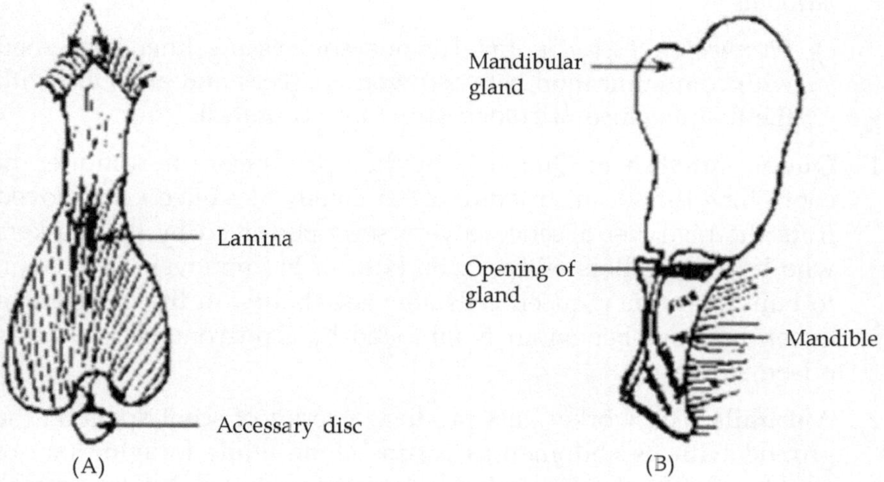

Fimbriae
Lamina
Accessary disc
(A)

Mandibular gland
Opening of gland
Mandible
(B)

(A) Aphrodisiac gland (butterfly) (B) Mandibles gland (honeybee)

b) *Sex attractant pheromones:* Pheromones employed by insects in bringing sexes together for mating are called sex attractant pheromones. They are secreated mostly by females then males.

Types of Insect Pheromone

1. **Pheromones attracting male insects:** Pheromone glands of female lie between posterior abdominal segments. The scent is released at

particular times of day which are the characteristics of the species. A sex attractants are commonly released after 1-2 days of emergence from the adult female and continue until a successful copulation. The effect of scent is to excits the male to initiate the mating. Males are seen copuling a cotton swab so could with the pheromone of the female. The Bombyxal is a sex attractant of *Bombyx mori* is an unsaturated alcohol.

2. **Pheromones attracting female insects:** The male insects produce pheromones to attract females (mecopatera). The pheromone is secreted from two vesicles near the posteror abdomnal tergite (dorsal portion of an articulate animal's gomite) on her arrival the male copulates with her and presents her with the remains of his prey.
3. **Pheromones attracting both sexes:** Both sexes are attracted in few insects. Example– Female of Dendroctonus and male of IPS and Lycus (beetles). The main function of the pheromone of lycus is to attract the population to form an aggragation near the flower of *melilotus.*

c) *Pheromones of social insects:* The pheromones are either, concerned with communication between workers (bees and ants) are with the maintenance of colony structure (Termites).

1. **Queen substance:** Queen honeybee produces pheromones in controlling the social structure of her colony. If a queen is removed from the hive, her absence is very soon perceived by the workers who became restless within ½ an hour of her removel. They begin to built emergency queen cells after few hours. In the presence of queen bee, this behaviour is inhibited by a pheromone known as queen substance.
2. **Ant trails:** The worker ants produce a series of scent spots on the ground with its abdomen as it runs along while foraging (act of searching of food), the scent is released mostly by Dufour's gland. But in certain ants it is released by poison glands. In ants several other pheromones are produced that induce gathering and settling of workers grooming other workers and food exchange.

d) *Materials secreting by glands involved in physiology*: The secreting of salivary glands accessory reproductive glands are involved in digestive and reproductive physiology of insects.

1. **Salivary glands:** The principle salivary glands in insects are maxillary glands and labial glands. The major function of these glands is to secrete digestive enzymes that help in the digestion of food materials.

a) *Maxillary glands:* These glands are found in collembala and propura in the larvae and certain Trichoptera these are also found in beetles.

b) *Labial glands:* Labial glands are paired structures generally situated in the thorax one either side of the foregut. These ducts combine to form median salivary duct which opens into the pre-oral food cavity near the base of hypopharynx. In cockroaches and grasshoppers they are commonly very large and composed of a number of lobes. Labial gland of Dorsophilla larva secret a mucoprotein used to cement the puparium to its substratium.

2. **Accessory reproductive glands:** In female and male insects there are generally 1 or 2 pairs of accessory reproductive glands. In locusta there are 15 pairs of accessory glands. These glands are associated with vasa differentia or ejaculatory ducts. However these are absent in silver fishes and flies.

ENDOCRINE GLANDS OR DUCT LESS-GLANDS IN INSECTS

These are duct less glands. Their secreations are called hormones. These hormones are transported to the target through blood. They act as chemical messanger's that bring about chemical coordination of the various parts of the body. The hormones of nervous system regulate the physiological and behavioural response of the insects. However on its action on physiological and behavioural responses is much slow that those mediated to nervous system. The endocrine glands are the brain, corporate, corporda cardiaca and prothoracic gland.

i) *Neurosecretory cells of the brain:* The pars intercerebralis of protocerebrum brain contains two grounds of neurosecretory cells that secreate ecdyriotropin and transport it to the corpus cardiacum from where it is subsequently released into the haemolymph. In lepidoptera it has been recently demonstrated that brain hormone is also stored in corpus allatum the brain hormone stimulates the secretory activity of the prothoracic gland and play important role in growth and metamorphosis.

ii) *Corpus allatum:* Corpora allata are glandular bodies, oral or elliptical in outline and lie behind the brain or in the neck on the side of oesophagus. They are usually paired but in but in higher dipterans they may be fused to a single body. The corpora allate produce Juvenile hormone that regulates the metamorphosis and yolk deposition in the eggs.

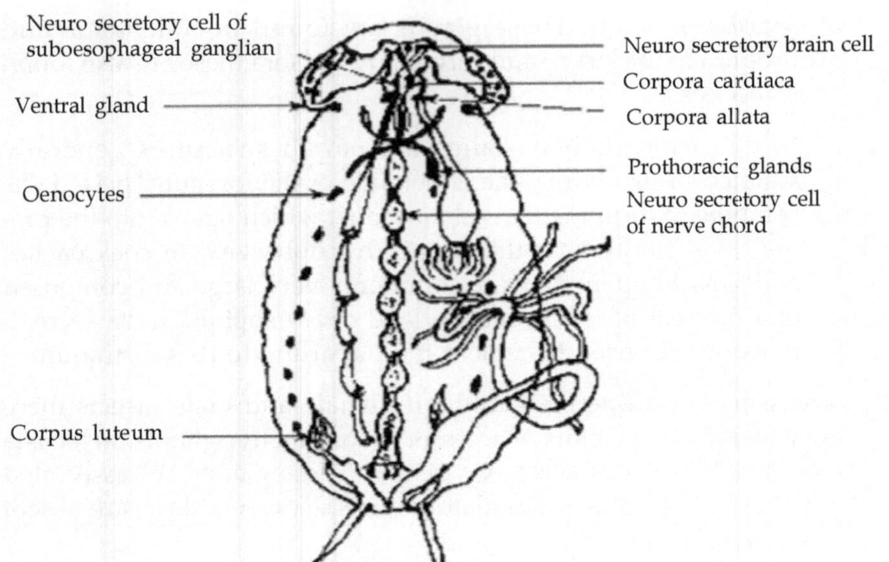

Secretions sites of the main hormones of insects

iii) *Corpus cardium:* The corpora cardiaca serve as neurohemal organs are a pair of small glands located behind the brain in close association with the aorta in addition to containing intrinsic secretory cells, the corpora cardirea receive axons from the neurosecreoty cells in the brain and serve as storage and release sites for their secreations. The intrensic secretory cells produce hormones which are concerned with the regulation of the heart beat.

iv) *Weisman's ring or Ring glands:* The small ring of tissue supported by trachea called Weismans ring is found in maggots of the higher Diptera. It is formed by the fusion of the corpora allata corpora cardiaca and the thoracic glands the ring gland is connected to the brain by a pair of nerves.

Significances of Hormones: The brain hormone or neurohormone is a peptide where as ecdysome is a steroid hormone and is probabily derived from cholestrol. The hormones act on genes through receptor mediated process determining which genes are brought into action at a given time, as a result influences the kinds of proteins synthesised. At least five non-sterolic compounds are found that act as juvenile hormone. The actual mode of action of hormone is not properly known but it is believed that they activate or repress target genes by binding through its association with a specific receptor to a regulatory sequence of the DNA. Like ecdysone, JH mimics are found in plants.

Example – Terpienofornesal

Hormones and neurohormone influence colour changes, osmoregulation, control of heart beat, amplitude, regulation of metabalic activities such as maintenances of carbohydrates levels in the hemolymph, synthesis of proteins and metabolism of lipids, control of seleropisation and melanisation of the cuticle, control of growth and metamorphosis, control of circadian rhythms, determination of sex receptivity, regulation of dominancy and regulation of migratory behaviour.

Prothoracic glands are diffuse, paired glands located at the back of the head or in the thorax. These glands secrets an ecdysteroid called ecdysone, or the moulting hormone, which initiates the epidermal moulting process.

11

General Ecology of Insect

The term ecology was first of all proposed by zoologist H. Pietir in 1865. It was derived from the Greek word ecologic means oikos = living place and logos = study and some definitions of ecology presented below:

i) **According to Ernst Haeckel, 1869** "The total relations of the animal to both its organic and inorganic environment is called as ecology".

ii) **E. Warming 1895** "Ecology is the study of animals and plants in relation to their environment".

iii) **According to Odum 1963** "Study of structures and functions of nature is defined as ecology".

iv) **Krebs 1985** says that the "Ecology is the scientific study of interaction that determine the distribution and abundance of organisms".

The life of insects is dependent on the adaptability of the environment, if the environment is unfavourable then the insect population *ceases* and *vice-versa*. This insect move one place to another to get the favourable environment so that they can pull on their life activities properly. There is a deep relationship between the living community and non-living environment. Both are affecting each other greatly. This inter-relationship between living community and non-living environment is called as ecosystem. The term ecosystem was first used by A.G. Tansley as early as 1935 and he defined, "Ecosystem is the system of resulting from the interaction of all the living and non-living factors of the environment".

Branches of Ecology

1. Autecology
2. Synecology

Autecology: The study of an individual species in relation to its environment is termed as autecology, therefore it is also known as ecology of individuals. To understand how an individuals fecundity, speed of development or expectation of the life is influenced by its environment are studied under this branch.

Synecology: The study of a species in relation to various other species, living in the same environment and constituting a community is termed as synecology. Synecology is divided in some type.

i) Population ecology

ii) Community ecology

iii) Ecosystem ecology

iv) Biome ecology

Components of Environment: Since insects inhabit, soil, water and air therefore the study of their environment is very complicated. Besides they live inside the plants as well as in the body of other animals (Parasites). Their growth survial, multiplication and distribution etc. are considerable affected by physical factors *e.g.*, Temperature, light humidity atmospheric pressure, wind, water, edaphic and by the biotic factors. According to Howard and Fiske 1911 analysed the environment into two set of factors whether the harmful effect of environment increased with density of population or was independent of the density which were latter called as "facultative factors" and "catastrophic factors". On the basis of above the environmental factors broadly may be grouped into two.

A) Abiotic of physical factors

B) Biotic or organic factors

A. Abiotic factors: Abiotic factors that directly or indirectly affect the population of insect includes temperature, moisture, light and serval other physical/chemical parameters. Weather is a composite condition of influence of temperature, light, humidity, rainfall and wind at any given moment in time. It varies continuously through out days, week's months and years and exerts an influence on insect abundance, distribution, longetively, development, rate so on from one years or season to the next.

Temperature: It is the most important factor of the environment. The Acting is directly on the development and metabolism of the organism which determines their abundances and indirectly through food and other environmental factors. The entire zone of temperature in which some development takes place is called as the zone of effective temperature and the minimum effective temperature is known as the threshold of development. The insects have different ranges of temperature at which they thrive. Some Diptera can survive at as high temperature as 55°C, whereas many grasshoppers and housefly thrive between 25°C and 35°C; Mosquitoes are more abundant in place having a temperature of 20 to 27°C, but not in places where temperature reaches 215°C.

Moisture: Most of the insects contain 80-90% water like other organisms the body composition of all insect consists of mostly water. Since water forms an essential constituent of the cells protoplasm, all the vital processes inside the body of an individual require water. One of the fundamental necessities of generally all the insects, therefore is to maintain a proper balance between the moisture contents of their body and that of their environment. This is achieved by regulating the intake and losses of water, with the help of a number of adaptations percent in insects. Thus optimal humidity varies for each species and for each stage in its life cycle. Extremes of environmental humidity directly influence the various activities of insects. Feeding development is reproduction etc. Under highly moist condition, silk worm larvae fail to pupa similarly newly emerged migratory locust live longer in the lower humidity. The desert locust lays its eggs in soil and 60% moisture content at 35.5° + 5°C temperature appears optimum for development low moisture content 3.7% of the soil gives rise to solitaria while higher content to gregaria phase. The number of hours of light in day length influences the development of aphids. The larvae of cutworm try to avoid sunlight by keeping themselves concealed during day but feed on the plants during darkness. All the insects are not equally active through out the 24 hours of the day.

Biotic factors

The biotic environment includes individuals of the same species or other species. The vital processes such as growth, nutrition and reproduction depend upon the interaction between the individuals of the same species or between those of different species. The interaction between species may be beneficial to both, harmful to both, to one and neutral to other. Therefore it is also considered one of the important biotic factors.

Interaction between organisms

It is two types:

i) *Intraspecific:* It consists of aggregation of individuals of the same species.

ii) *Aggregation:* Aggregation means the tendency of some insects to concentrate.

Light: The requirement of light varies from species to species and even one stage to another during development of insect. The light is an essential ecological factor for various biological processes. Silkworm develops faster in light than the darkness, similarly the grubs of Trogodema granarium develop more rapidly in light than the darkness and their breeding under constant light adversely affects the amount of oviposition of the resulting adults as compared to darkness. The moths of spotted boll worm of cotton and of the red hair caterpillar lay most of their eggs during period of darkness. There is correlation between the activities of honey bees and hours of sun skins. The paddy leaf hopper is attracted to light on a hot and humid evening but is indifferent to it during dry weather, the temperature remaining the same. The diapause is initiated by day length of shorter duration. Photoperiod in addition determines whether it is embryonic diapause, larval diapause, and pupal diapause in number larger than found in normal distribution and congregates under various circumstances mostly by their own movement.

iii) *Association of the insect sex:* Sex ratio of insect plays a vital role in the build up of their population.

iv) *Parental care in insect:* It is well known fact that all insect take care that the eggs are laid in situation where condition are favourable their hatching as well as presence of sufficient food for newly young ones. There are some insects which lay their eggs in protected places or cover with foreign materials or prepare brood nests to protect them from natural enemies.

v) *Social life of insect:* The honeybee's termite's ants and wasps have developed social life which is good for the coloney members. The relationship is not for only living together in large member but it also involves the division of labour among the members for the welfare of the community as a whole.

Predators of Insects

If predators feed on the one species of prey, it is known as obligate predators and if feed on more one species than it is calle as "Herpectors marginatus" predators of some species *Chrysocoris stolli* and aphids; when one insect attacks an individual one or more species of insect of obtaining its food.

Parasitism of Insect: When an organism lives in or on the body of another organism and feeds at the expense during it entire feeding periods the association is termed as parasitism. The feeding individual is known as parasite and the individual on whom it feeds is as the host.

INSECT ECOLOGY OF UNIT

Introduction

Insects are the dominant group of organism on earth in terms of both taxonomic as well as ecological function. The lives of insects are interwined with our own life. Insect interaction with the environment is not due to their occurence but their vast diversity. The diversity of insect species represents an equivalent variety of adaptation to the variable environmental conditions. Insect ecology explains the dynamics of insect number in time and space. Study of insect ecology is important because it provides insights into insect evolution and enables to develop ways.

Insect play an important role in ecosystem functions as they represent food resources, predator's disease vector for many other organisms. They have capacity to alter rates and direction of energy. Insect ecology is the study of interaction between insects and their environment.

SCOPE OF INSECT ECOLOGY

It is a multidisciplinary subject.

Ecosystem Conditions

Climate

Substrate

Energy flow

Biogeochemical cycling

Individual traits

Tolerance

Range Mobility Resources

Community structure

Population system

INTERACTION BETWEEN ECOLOGICAL ORGANIZATION

It is the integrative field of study requiring the contribution of biologist chemist geologists and other to fully understand the interaction between them.

ENVIRONMENT AND ITS COMPONENTS

Ecosystem is the basic functional unit of environment Interaction b/w the living and non-living components of the environment. Environment of an insect population consists of two factors:

1. Physical factors (Abiotic factors)
2. Biological factors (Biotic factors)

Abiotic factors

Temperature

Wind: climatic factors Humidity

- Light
- Medium (soil) edaphic factors

Biotic factors

Members of the same species (Homotyphal effect) Members of other species (Heterotyphal effects) Nourishment, food source.

Anthropogenic factors: Impacts of human activities

ABIOTIC FACTORS OF INSECTS

A biotic factors such as temperature moisture, light and several other physiochemical parameters directly or indirectly influence the insect population. The biotic factors act not only independently but also collectively. Combined effect of these factors is referred as weather. Effects

A) **Temperature:** Insects are basically poikilothermic and their temperature is variable. They have little physiological regulation of body temperature. However are they posse's behavioural adaptation that maintains the body temperature as near to optimuna temperature? An Insect may survive high or low temperature during certain stages of life cycle. *e.g.* many insects are able to survive much lower temperature in winter than in the summer. Tropical insects are generlly less tolerant of cold than those in temperature regions optimal temperature range for most of insects is 22° to 38°C. Some species are able to survive at temperature beyond these ranges. *e.g.* some dipteran larvae may survive at 55°C or even higher while certain beetles may develop at 0°C

   a) *Lower lethal temperature:* It is the temperature below the optimum range at which insect become less active. It exposure period is prolonged they die due to starvation. *e.g. Locust* stop feeding below 20°C. At temperature below freezing point majority of insects die.

   b) *Upper lethal temperature:* It is the lethal temperature above the optimum range of temperature at which activity of insect increases.

   c) *Acclimatisation of low and high lethal temperature:* The responses of temperature are continous and vary according to experience of the insect. This is known as acclimatisation *e.g.* larvae of mosquito reared at 30°C, die at 0.5°C and survive at 18°C to 20°C.

B) **Humidity or Moisture:** Humidity or moisture or water vapour content in the environment directly or indirectly affects the reproduction.

a) *Water content in the insect body:* The water content of insects varies from less than 50% to more than 90% of total body weight. As mentiond for temperature variation occur both among different species and among different life stages. Soft bodies insects (*e.g.* caterpillars) have large amount of water where as many hard bodied. Insects have thick certicle *e.g.* (beetle) have less amount of Water Active stages commonly have higher water content than dormant stage.

Insect such as silver fish able to absorb the moisture is directly from atmosphere. Some are able to maintain the water content of body.

b) *Environmental moisture:* The precipitation (*e.g.* Rainfall snow, hail and sleet), condensation (*e.g.* dew, fog, white frost) and available surface water are common forms of water available in the environment.

Humidity in air depends upon temperature and atmospheric pressure the relative humidity varies with location, time of day or year, topography vegetation.

Rate of development is also affected by humidity. The incubation period of *Ptinus* eggs is 15 days at 20°C and 30% RH but is 10 days at 90% RH at the same temperature. Rate of development in *Locusta* is tastest at 70% RH. Heavy rainfall cause mechanical damage to insects. (*e.g.* Aphids).

C) **Light:** Light has environmental influence rather than as survival factors. It includes photoperiod, intensity and wavelength. They do not cause mortality.

a) *Photoperiod:* It is the measure stimuli that induces diapause long day (photoperiod more than 12 hours) occur in summer short day (photoperiod less than 12 hours) in winter with increasing and decreasing day lengths. *e.g.*, The photoperiod less than 16 hours inducs diapause in the pupae of *Acronycta.* Photoperiod more than 14 hours initates diapause in Bombyx mori.

b) *Light Intensity:* The light intensity is the major factor that control the flight behaviour in Insects (*e.g.* Butterflies, wasp and bees) are not active in dark and aphids do not fly and low intensity. Nocturnal insects *e.g.* (moths) do not fly in light. Effect of light on both aquatic and terrestrial plants may indirectly influence the activities of Insects.

c) *Wavelengths:* Phytophagous insects locate their food plants by using different wavelength of reflected light. The orientation and navigation of insects are guided by position of sun and degree of polarisation of light in different parts of sky *e.g.* Diurnal crickets posses the ability to use polarised light in returning home.

D) **Air and Water current:** These two are effective agents in distribution on insects. *e.g.* Aphids, leaf hoppers are blown thousand of miles by air current. It is modified by vegetation water current is influenced by volume of water, slope of stream bed and temperature Airmovement cause mortalily in two ways:

1. Severe wind 2. Heavy rain

Migrating insects use rising air current (thermals) by soaring and thus conserving energy *e.g.* monarch butterfly water current often determines the habitat of certain insects, *e.g.* may flies (Ephemeroptera) with clinging legs and streamlined body are associated with fast moving water. Many aquatic insects are unable to survive in moving water, *e.g.* mosquito larvae. Biotic factors

It depends upon the interactions that are found in insects. There are basically two types of interactions:

Intraspecific Interaction

Interspecific Biotic Interaction

1. **Intraspecific Interactions:** It consists of aggregation of individual of the same species. It consists of Aggregation, canabalism Association of sexes, parental care. It involves beneficial effects *e.g.* large aggregation of individual by their overall appearance, deter predation.

This interaction with high population density may be advantageous in mate finding, survival and potntial predation *e.g.* this interaction enhance predation in assassin bug, Rhynocoris longiferous. Low population density reduces the probability of finding mate, lower average fecundity. At limited resources these competitions lead to death. Thus it regulate population size (Intraspecific interaction).

2. **Interspecific Interaction:** Different species of insects and their interaction. The major interation found in insects are competition symbiosis, predator-pray interaction etc.

Odum (1971) describes all type of relationships between organisms into two: postive interations and negative interactions.

1. **Competition:** It is a negative interaction. This interaction arises when needs of two or more different species for a given source (food or shelter etc.) coincide with their overlapping niches.

Two species of identical needs competes one displaes the other. This is known as competitive displacement *e.g.* two species of flour beetle, *Tribolium costaneum* and *T. confusuns* are placed in a container at high humidity and temperature, ***T. castaneum*** displaces ***T. confusum.***

2. **Symbiosis:** Close association between two different species. Symbiosis may be advantageous (*mutualism*) disadvantage for one species (parasitism advantage for one without doing harm to other (*commensalism*).

   a) *Mutualistic association:* It occurs in insects *e.g.* ant cares and protect aphid which in turn secrete honey due (a solution produce from anus) *e.g.* Assassian bug, lophocephalla and crazy ant *Anopiolepis.* The ants accomate the bug in their nests, guide them to feeding grounds, protect them from predators while the bugs feed.

   b) *Commensalism:* It also exists among insects. Many microbes associated with insects help their hosts and gain food and a palace to live. Phoretic relationships are good example. Insects may act as transporters or riders or both *e.g.* chewing lice (Mallophaga), the ectoparasite of birds attached by their mouth parts to louse flies are transported from one host to other.

   c) *Parasitism:* Some insect may act as parasite or act as vectors. Disease carrying pathogens are carried by the insects. The role of microbes in biological control of insect pest is an important aspect of research. Many bacterial species are identified as inset pathogen. Pathogenicity of insect is usually due to production of host-specific toxins. *E.g.* Best known bacteria that infect insects is *Bacillus thuringiensis.* It is used against mesquito larvae. Are passed maternally and kill embryonic male flies thus affecting the sex ratio.

3. **Predation:** The predator of insect's includs insects and other arthopods such as mites, spider etc. Human beings by them are interation with insect's exists a major impact on the insets. It includes modern agriculture, urbanization, transportation, impact of insecticides and so on.

The effects of air, water pollution on the environment give further way to this interaction.

INSECT PEST

Insects are one of the most successful groups of animals and therefore they are the greatest competitors of man in the struggle for existance. "All insects that cause significant and economic damage of 5 percent or more to the crops, stored products, poultry birds and domestic animals are called pests." If an insect causes a loss of less than 5 percent, the infestation is said as negligible. Insets which normally cause a loss ranging from 5-10% are minor pest and those which cause a loss of 10% or minor pest and those which cause a loss of 10% or more of yield are called major pests. Various terms are used to differentiate the pests:

i) **Regular pests:** They occur most frequently on the crops and such insects have close association with particular crop. *E.g.* fruit borer on brinjal and lady's fingers, borers of sugarcane, hoppers on mango etc.

ii) **Occasional pests:** They rather occur infrequently and close association with particular crop is absent. *E.g. Heliothis* on wheat, mango stems borers etc.

iii) **Seasonal pests:** The insects which occurs a crop mostly during a particular part of the year are called seasonal pest. The incidence of this pest is largely governed by the climatic and weather conditions in a particular area *e.g.* hairy caterpillar on cotton.

iv) **Persistent pests:** Insects which occur on a crop almost throughout the year. *E.g.* mealy bugs and scale, termites etc.

v) **Sporadic pests:** They occur in few isolated localities and their population remain low for long period of time *e.g.* paddy root weevil. On the basis of feeding habit they may be classified as:

   a) *Monophagous insects*: Confined to a single species of plants and feed on a group of closely related plants *e.g.* monophagous insect in mulberry silk worm.

b) *Oligophagous insects*: Feed on a group of botanically related plants usually within a single plant family *e.g.* Potato tuber moth which attacks only potato mustard sawfly is also confined to cruciferae.

c) *Polyphagous insects*: Feed on many plants from a diverse range of plant family's *e.g.* locusts, grasshoppers, termites, hairy caterpillars, gram pod borer etc. Causes that make the insect as pest

i) **Invasion:** Promotes the exchange of perishable as well as other agricultural commodities, however, the increased trade results an increased risk of new pest invasion. Large proportions of the major pests in a modern agro-ecosystem are exotic in origin, *e.g.* Introduction of San Jose scale and cottony custion scale of citrus into India.

ii) **Ecological changes:** Various agrotechnical practices like, monoculture, selection of high yielding plant cultivars, or elimination of natural enemies etc. created conditions favourable to certain insect species and thus induced increase in their population. Whenever, the interaction between phytophagous and entomophages is disrupted, the population of the phytophagous insects increased tremendously and they attain pest status because they become free from the constrains imposed by them.

iii) **Socio-economic changes:** The human activities have provided condition conductive to breeding of mosquitoes and use of insecticides by man has created a number of secondary insect pests. Introduction of susceptible and nutritious host plants into the environment of a pest specis usually lead to its fast development and abondance. The insects feeding on wild plants don't occur in vast number. However, contrast to this endemic situation, the insect population is extensively enormous in the cultivated filds (epidemic situation) *e.g.* cotton jassid was not so serious a problem on the indigenous cotton in India, but with the introduction of American cotton, this insect became the dominant pest of this crop in central and South India.

BIOTIC POTENTIAL AND ENVIRONMENTAL RESISTANCE OF INSECT

Biotic Potential: It is described by the unrestricted growth of population resulting in the maximum growth of that population, Biotic potential in the highest possible vital index of a species so, when the species has its highest birthrate and lowest mortality rate.

Quantitative Expression: The biotic potential is the quantiative expression of the ability of a species to face selection in any environment. The main equilibrium of a particular population is described by the equation.

Biotic potential

No. of individuals = Resistance of the environment (Biotic and abiotic)

Resistance of the environment (Biotic and abiotic)

Champan also relates to a "Vital Index" regarding a ratio to find the rate of surviving members of a species.

Number of births

Vital index = Number of deaths × 100

Components Reproductive potential (Potential nativity) is the upper limit to biotic potential in the absence of mortality.

Survival potential: It is the reciprocal of mortality.

Reproductive potential doesn't account for the number of gametes surviving; survival potential is a necessary component of biotic potential.

In the absence of mortality, biotic potential = reproductive potential

Additional components of survival potential given by chapman are:

a) **Nutritive potential:** It is the ability to acquire and use food for growth and energy.

b) **Protective potential:** Ability of the organism to protect itself against the dynamic forces of environment in order to insure successful reproduction and offspring. A species reaching its biotic potential would exhibit exponential population growth and be said to have a high fertility, that is now many offsprings are produced per mother. Environmental resistance factors are all the things that keep a population of organisms from endlessly increasing. They lower the chances for reproduction, affect the health of organisms and raise the death rate of the population. Environmental resistance factors include factors that are biotic and abiotic.

Biotic factors: It includes things like predation, parasitism, lack of food, competition with other organisms and disease.

Abiotic factors: It includes drought, fire, temperature and even wrong amount of sunshine.

THE PUSH AND PULL OF POPULATIONS OF INSECTS

While environmental resistance acts like a hill pushing back against population growth, biotic potential is what urges a population to grow. So, while the biotic potential of a species causes the population to increase, environmental resistance keeps it from increasing relentlessly. When the population is small, environmental resistance factors are well, not as big of a factor. There may be plenty of resources around so the population can keep growing quickly. But, as competition get stiffer and resources start to become limited, population growth starts to slow. When the birth rate and death rate are the same, the population has reached equilibrium. For that population, in that particular environment the *carrying capacity* has been reached. No more individuals can fit.

Growth curve

How biotic potential and environmental resistance combine to cause a population to graw or decline.

1. **J-curve:** Before environmental resistance factors kick in to reduce population size.
2. **S-curve:** A balance between environmental resistance and biotic potential.

CAUSE FOR OUTBREAK OF PEST IN AGRO-ECOSYSTEM

Insect pests an insect whose population increases to the extent that it starts causing inconvenience, annoyance to man, animals, plants and material possessions.

The increase in the population of species of insects is known as *pest outbreak.*

Various causes for outbreak of pest in agro-ecosystem

1. Rapid increase in reproduction potential of an insect.
2. Increase in fecundity *i.e* laying more number of eggs by female.
3. Female population higher than that of male population.
4. Easy availability of sufficient quantity of the food for insect development.
5. Sometimes the life-cycle of an insect is shortened with the result many generations are completed in a year.

The various other factors are also responsible for the outbreak of pest, these are as follows:

a) *Deforestation:* Many insects are having their habitat on grass and forest trees but due to deforestation they have to migrate on cultivated crops in the absense of their enemies, which stimulate the growth of insects.

b) *Favorable climate:* The favorable climate leads to the rapid multiplication, those conditions whicch are unfavourable for its natural enemies give good chance for the growth and multiplication of insects.

c) *Destroying natural enemies:* The well known natural enemies of insects like snakes, frogs, birds etc. eat different types of insects and helps in maintaining natural balance. If such natural enemies are destroyed, the natural balance is disturbed and it will lead to increase in insect population. *E.g.* During 1956, a large number of sparraws were killed in china which result the increase of insect pests from the second year. It forced to the Govt. of China to use insecticides in large quantity during 1959.

d) *New pest introduction:* Sometimes, an insect is introduced into a new area, without its natural enemies, allowing it to breed n an unrestricted manner as a result it soon takes the status of the pest. *E.g.* Potato tuber moth came in India from Italy with potatoes and hence becomes a regular pest.

Modern Agronomic Practices

Timely irrigation, mechanical tillag, synthetic fertilisers improve the growth of the crop, reduce competition for food for insects, it leads to their fast reproduction. *E.g.* incidence ofstem borers in rice and aphids in cotton increases when high amount of urea is applied.Intensive and Extensive Cultivation of Crops. When one or more crops are cultivated in extensive areas, due to high food availability, there is no problem for the insect for food and shelter. The effect is more pronounced if cropping is done in more than one season for the year. If the different crops in rotation are closely related to each other or when there are afternative food plants for the insect concerned, the population of insect is likely to increase. *e.g.* incidence of borers and leaf hoppers when sugarcane crop is raised over extensive areas. Resurgence of Sucking Pests Plants treated with systemic and contact insecticides though afford initial protection against sucking insects tend to offer physiological conditions favourable for their rapid reproduction. *E.g.* In recent years in northeastern U.P brown plant hopper resurge on paddy crop due toheavy use of contact poisons against stink bug. Frequent use of Pesticides Indiscriminate use of pesticides leads to an outbreak of insect in 3 ways.

i) It kills the natural enemies.

ii) It kills non-targeted insects and removes competition and struggle for existance.

iii) It causes selective eradication of insects susceptible to insecticides so as to leave only the resistant ones to breed unaffected.

BALANCE OF LIFE IN NATURE

The concept of life system in Insects: A life system is the part of an ecosystem which determines the existance, abundance and evolution of a particular population (Clark *et al.*, 1967). Hence, the life system of an insect includes the part of environment *i.e effective environment* which directly influences the fate of a given population and the components of the population.

Like forest and pond ecosystem, large crop areas are maintained to satisfy human need and these are called the *agroecosystem.* It's an artificial human engineering ecosystem.

In an agroecosystem, the landscape is altered, the original natural vegetation has almost entirely been removed and profitable crop plant is introduced.

Big trees are eliminated resulting into loss of bark; canopy etc *i.e.* many microhabitats of insects. A typical agroecosystem is usually composed of the more or less uniform crop-plant population, weed communities and the physical environment with which they interact.

About 30% of the terrestrial world is devoted to agriculture in the broadest sense. Major portion of the land mass in agriculture areas is occupied by the agroecosystems.

When an agroecosystem is improperly managed there is every possibilitity of the same to resume back to natural ecosystem. In natural ecosystems, a state of balance exists or will be reached *i.e.* species interact with each other and with their physical environment in such a way that, on average, individuals can only replace themselves. As results, each specis in the community achieve a certain status that becomes fixed for a period of time and is resistant to change. This phenomenan is referred to as the *balance of nature.*

Agricultural practices with monoculture aiming or full reproductive potential causes a stress. At the same time strong forces react in opposite direction to the imposed changes (*i.e.* natural ecosystem to agroecosystem) towards its return to the original system (*i.e.*

agroecosystem to natural ecosystem). An insect outbreak may be initiated as one of these forces. Destruction of crops by insects is a force against the disturbing factors *i.e* overpopulation by the crop species. So, pests are not ecological aberrations. They function like other animal populations. Their only distinct in this regard is that their activities usually counteract the wants and needs of human populations (Pedigo, 1996). Both biotic and abiotic components significantly influence the insect population like any other population. When one of these components approaches or exceeds the specific limit of tolerance of the species, relative to survival, development, fecundity etc. it becomes a limiting factors. Thus, it limits the extent of survival, degree or rate of development, fecundity etc.

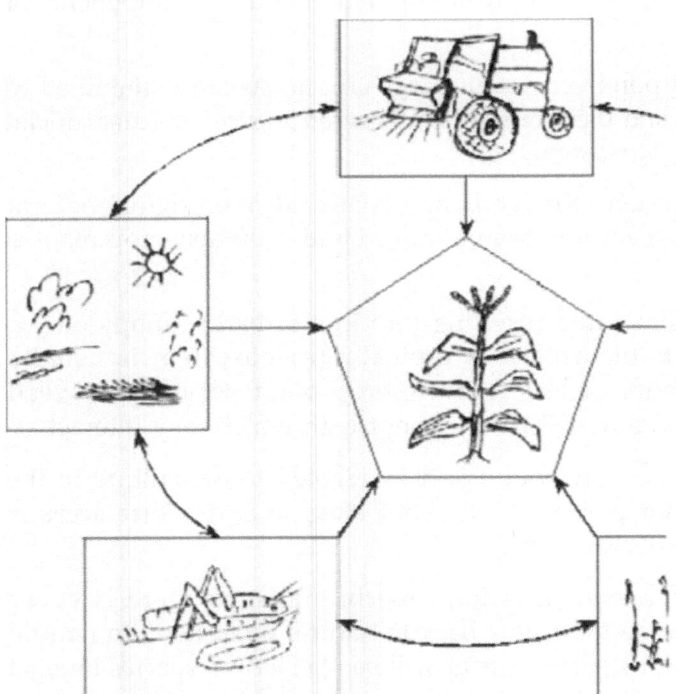

Major interacting elements of agro-ecosystem
(Human inputs. Physical environment, Insect, Crop)

Species in the community achieve a certain status that becomes fixed for a period of time and is resistant to change. This phenomenan is referred to as the *balance of nature.* Agricultural practices with monoculture aiming or full reproductive potential causes a stress. At the same time strong forces react in opposite direction to the imposed changes (*i.e.*

natural ecosystem to agroecosystem) towards its return to the original system (*i.e.* agroecosystem to natural ecosystem). An insect outbreak may be initiated as one of these forces. Destruction of crops by insects is a force against the disturbing factors *i.e* overpopulation by the crop species. So, pests are not ecological aberrations. They function like other animal populations. Their only distinct in this regard is that their activities usually counteract the wants and needs of human populations (Pedigo, 1996). Both biotic and abiotic components significantly influence the insect population like any other population. When one of these components approaches or exceeds the specific limit of tolerance of the species, relative to survival, development, fecundity etc. it becomes a limiting factors. Thus, it limits the extent of survival, degree or rate of development, fecundity etc.

The environmental components do not uniformally exist throughout an ecosystem. For example, the temp of one part of the plant may differ from that of the other part. This leads to the development of the concept of *microenvironment* which implies the recognition of spatial and temporal differences in environmental components within an ecosystem. The population studies are aimed to understand the composition and the behaviour of populations, the impact of environmental determinants on population and ultimately to develop predictions about these populations. Populations have characteristics that can be measured and described. They are as follows: genetic composition (*i.e.* genotypic and phenotypic variations), sex ratio (*i.e.* the female to male proportion), *age composition* (*i.e.* composition of adults and immatures of varying ages) etc. The other characteristics are dispersion (*i.e.* arrangement of individuals in space according to the availability of resources) *population size* (*i.e* total no. of individuals), *population density* (*i.e.* no. of individuals per unit area or volume), biomass (*i.e* total weight of populations and dynamics) (*i.e* changes either in no. or density or both over a period of time).

The inherited properties (*i.e* genetic variations) affect the ability of reproduction and survival. The reproductive ability varies among insects and is known s *potential natality,* (*i.e* reproductive rate of individuals in an optimal environment. *Survival rate* also changes among insects and depends on feeding habits and protection of young ones. The insects feeding on a wide variety of food resources and procecting their young ones to a certain degree have greater survival potential. Viviparous and ovoviviparous insects have generally lesser mortality of their new borns than in oviparous insects. The insects with high potential notabilities tend to have relatively low survival rates and *vice-versa*. The insects with high reproductive rates and low survival rates can rebound quickly after environmental adversities and are termed *r-strategists e.g.* aphids. On the other hand, the insects with low reproductive rates and high survival rate compete effectively for environmental resources and their survival rate and are termed k-strategists *e.g.* codling moth.

The capacity of an organism to increase exponentially under suitable conditions is referred to as the biotic potential. However, biotic potential can never be fully realized under natural circumstances due to environmental factors like predation, limitation of resources etc. which prevent such increase. The whole collection of these factors is called *environmental resistances.* The dynamics of a given population is the outcome of interaction of the biotic potential and the environmental resistance. The dynamics of a population can be studied by the following three basic approaches.

1. Analysis of population under controlled but artificial environment.
2. Evaluation of population in the field.
3. Theoretical or mathematical models that describe population dynamics.

12

Insect Collection, Preservation and Insects Control Methods

Many insect can be observed at any hour of the day. This is proper understanding of the taxonomy and biology. The insect collection, storage and preservation are useful technique of their culture. The best period for collection is from early spring to late rainfall and the best time of the collection of most of the species is during the daytime. The insect are polyphagous and therefore, plants provide one of the best collecting places. Insect can be picked, swept off the plant with the help of a net. Different species feed on different kinds of plant, one should therefore, examine all sorts of plant, flowers weed shrubs and trees. Few insect are only found in the stem, bark, wood, fruit or roots etc. Some species can be found in the leaf mud and litter on the surface of the soil, particularly in woods or areas where the vegetation is dense, others can be found under stones bark etc. other house hold insects may be observed feeding on clothing, furniture, grain, food and other parts of Plant and House and some are aquatic insects.

INSECTS COLLECTION AND PRESERVATION OF INSECT METHODS

1. **Hand Picking Methods of Insect:** Some minute insect and soft bodied ones should be collected by hand either with the help of a fine forcep or brush. The soft brush should be dipped into the medium in which the insect have to be preserved so as to minimise the damage to soft skin of Insect. Insects like leaf miners, order Diptera, Aphids, Hemiptera, bark beetle insect, Coleoptera order also collected by hand picking method.
2. **Aerial Netting Collection of Methods:** The aerial net are most widely used to collect free living fly insect *e.g.*, Moth, Dragonflies, Butterflies, Honeybees, Wasp yellow etc. After netting the insect, the net should be turned to prevent the escape of captured insect,

the soft bodies insect like moth may be gently removed from the bottom of the bag. 3. Collecting with Aspirator Methods of Insect. The insect like leaf hoppers, family pentatomidae and order colleoptera etc. may be collected by a sucking tube or aspirator straight from the plant surface. It is a simple device and if used with little patience and caution may yield desirable result. A vacuum cleaner may also be used as aspirator to suck the insects from herbage.

TRAPPING METHODS OF INSECT

Trapping is a easiest method of collection of insect. Some common types of trapping methods are (1) light trap (2) bait trap (3) wind trap.

i) Light trap by artificial light like kerosene lamp, petromax gas light, electrical lamp.

ii) Bait like over ripe fruit, Piece of meat, meat of fish, rotten-fungi and animal excreta. Put at proper place where insect many attract.

iii) Wind trap is a insect trapping method/where insect is traped in the direction of wind it may be electrically operated suction device.

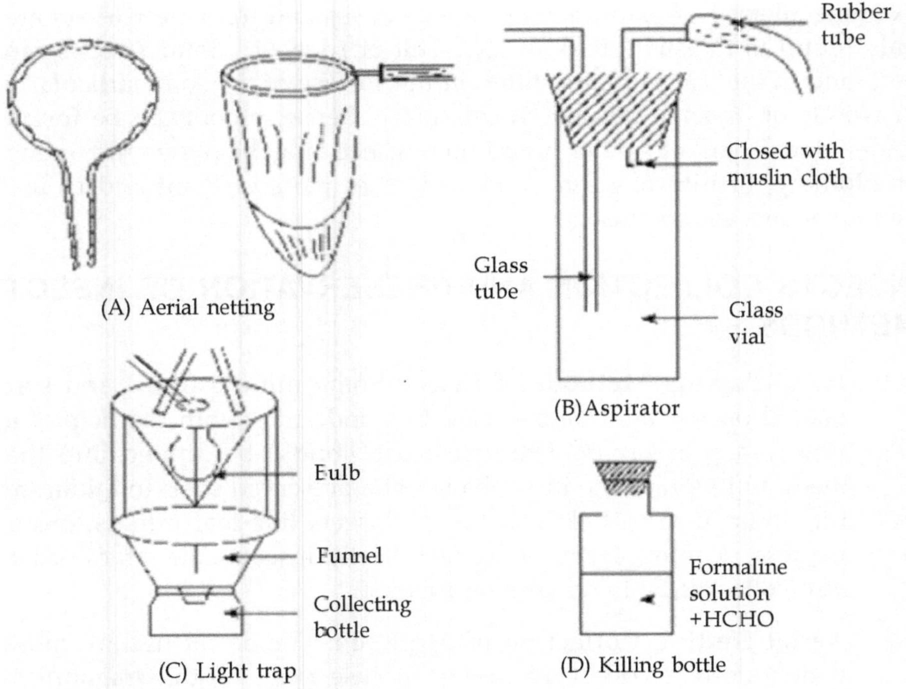

(A) Aerial netting, (B) Aspirator, (C) Light Trap, (D) Killing Bottle

Insect collecting equipments

1. Net
2. Aerial nets
3. Sweep net
4. Aquatic dipnets
5. Brush, Forcep scissors
6. Aspirator
7. Knife, Axe, hammer
8. Killing bottle
9. Collection vials
10. Hend lens
11. Paper Packets and envelopes
12. Cotton and chemicals
13. Traps
14. White tray and sieve tray
15. Camera and boot
16. Field notebook
17. Lunch packet

Collection Mounting and Preservation of Insect

First collect the insect then mount it and finally preserve for the further examination of the specimens for identification and study of insect. To small insect first pin and dry and will keep indifinitely it pinned by minute pins or on microscopic slids. And for large insect like butterflies, moth, grasshoppers etc. may be mounted in various types of glass topped display cases, soft bodies forms such as nymphs and larvae like insect should be preserved in fluidslike Alcohols (means 70 ml Alcohol and 30 ml distilled water).

Pinning of Insect: The pinning is best way to preserve hard-bodied insect. Pinned specimens keep well retain their normality and are essay to handle and study. Insect are usually pinned vertically through the body as shown in figures *e.g.*, bugs beetles moth are pinned through the thorax center part the body. All specimens should be mounted at a uniform high on the pin about 20 cm above the point. Hemiptera bugs are pinned through the scutellum, a little of the right of the mide line if the insect suitellum is large *e.g.*, all bugs after the collection, the insect should be mounted as soon as possible as after dying the body become brittle and may be broken in the process of being mounted specimen stored in envelopes for a long time must be relaxed before being mounted. Usually most of the insect are relaxed to mount after two day in such a chamber.

Spreading Insects: The wings of an insect is spread on a spreading broad, dorsal side up and the pin left and right side in the insect. There are certain standard positions for the wings to spread insects. In the case of moth and housefly the rear margins of forewing (attached thorax) should be straight across, at right angles (90°) to the body and hind wings should be far enough forward that there is no large gap in pin at the side between the fore hind wings. In correctly spread wings is the biggest error in beginning entomology collections. Proper spreading takes practices. Poorly spread specimen should be replaced before entering collections in competition.

Mounting Methods of Insect

The procedure of mounting a specimen on a microscopic slid material depends on the group of the insects and on the type of mounting medium used. Dehydration involves running the specimen successively through increasing concentration of alcohol series 30%, 50%, 70%, Acetocarmin staining material or Borax carmine and 90%, 100% solute alcohols and finally through xylene and finally into the canada balsam/DPX on the slids the time left in each solution before moving it to the next depends on the size of the insect and may vary from a few minutes to an hours. Inside the DPX the insect is oriented and then cover glass is put on. Once the cover glass is on the slid must be kept horizontal untill the balsam hardens. This is involved in microscopic animals, mosquito larva and protozoan's animals. The process of permanent mounting involves a series of treatment.

Precautions

1. The slide, cover slip, needle, brush etc. should be clean.
2. The material should be in the centre of cover-slip.
3. There should be no air bubble in the conada balsan or DPX.

Labelling and Identifications of Insect

1. A specimen labelling insect
2. Name of the host and location
3. Locality district
4. Date of collection
5. Name of the collector
6. File cover papers

Identification of insect always demands well-preserved specimens and reference collection to compare with order to order insect. Useful taxonomic literature along with reference collection on the group can be said to be the keystone for long term entomological work. In the beginning one has to build up a reference collection normally getting the identification and getting the confirmation of tentative identification done by specialist working in museum and institutes.

The photographs of insect are provided to assist student in basic insect identification. Be aware that dichotomous keys, similar to those used for order classification, also exist for separating insects into families and even further into genera and species. The use of such keys is beyond the scope of this book in most cases.

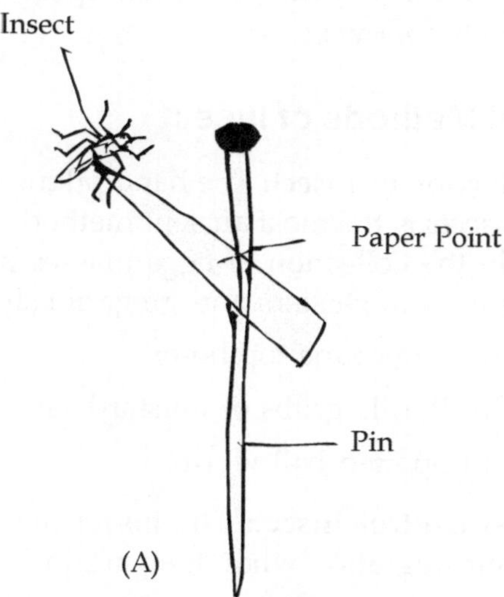

Mounting of a minute bug on Paper Point Pin

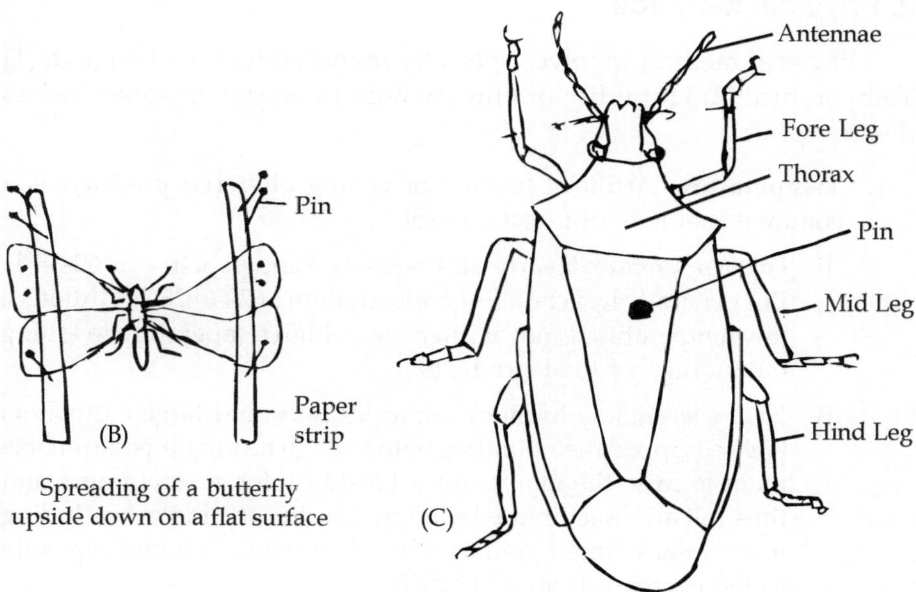

Methods of Pinning of insects in bugs

THE INSECT CONTROL METHODS

Insect control in broader sense may be defined as to make the life of an insect harder or to prevent their increase or spread or to kill them by any means. It depends upon the knowledge of the factors causing the abundance of the particular species. The insect control of the some methods follows :

1. Mechanical Methods of Insect

1. **Hand collection of insect:** The hand piking of large sized immature stage of insects is the most ancient method and this may prove very effective in the collection of eggs masses and larva of such insect which lay eggs in clusters and gregariously.

   e.g., i) Sugarcane top borer

   ii) Pyrilla grubs of mustard saw fly

   iii) Spotted ball worm.

2. **Hand nest control insect:** The insect may be caught very easily during their migration when they are in large number *e.g.,* pyrilla, moths etc.

2. Physical Methods

Physical method involves specially manipulation of change in (1) Temperature (2) Humidity or employ radiant energy in some way to destroy pest.

1. **Temperature:** Artificial heating or colling of stored products is a common methods of insect control.

   i) *Low temperature:* Nearly all insects become in active at 60-38°F. But generally insect undergo hibernation at 20 to 28°F. Although low temperature is not as effective as high temperature in killing but storage of food products.

   ii) *High temperature:* In many countries mills and large cultivators have equipped their building with enough heating pipes to enable them to raise the temperature 120-145°F for several hours and thus kill all insect in the building. Most of the insects including insect attack stored grain can be killed within 2 hours exposure to temperature from 120-128°F.

   iii) *Humidity:* The draining of marshy land water is the most effective methods or destroying all sucking insect and sponge insects. Some crop can be protected from insect pests by irrigation at proper time of day.

iv) *Light:* The light has been utilized to attract many strongly phototrophic insects into traps from which they can not escape. The use of radiant energy to control insects has been used in various laboratory experiments, studies etc. by a number of workers. The light is a most important factors of day. The day sterile male released in field should not damage crops. Generally insects appear to be strongly attracted at about 2210A hence different type of lamps have been used in some countries to trap the flying moths.

3. Biological Control of Insect

Biological control is the use of natural enemies:

- Predators
- Parasites
- Pathogens and competitors to control. Pests and their damage, plant pathogens nematodes and many natural enemies.

Predators. Predators are mainly free living species that directly consume a large number of prey during their whole lifetime.

Ladybugs Beetle and their larvae which are active between May and July in the northen hemisphere, are voracious predators of *aphids* and will also consume *mites, scale insect* and small caterpillars.

Dragonflies are important predators of mosquitoes, both in the water where the dragonfly *naiads* eat *mosquito larvae* and in the air, where adult dragonflies capture and eat adult mosquitoes.

The community wide mosquito control programes that spray adult mosquitoes also kill dragonflies, thus act as important biocontrol agent.

Other useful gradan predators include *lacewings, pirate bugs* rove and ground beetles *aphid midge centipedes spiders predatory mites* as well as larger found such as *frogs toads, lizards hedgehogs, slow-worm* and birds. Gate and rate terriers kill field mice, June bugs and birds.

Parasitoid insects. Parasitoids lay their eggs on or in the body of an insect host, which is then used as a food for developing larvae. The host is ultimately killed most insect parasitoids are wasps or flies and usually have a very narrow host range.

Tachinid flies. Parasitize a wide range of insects including caterpillars, adult and larval beetles true bugs and others. Parasitoids are one of the most widely used biological control agents commercially there are two types of rearing system short-term daily output with high production of parasitoids per day, and long term low daily output with a range in production of 4-1000 million female parasitoids per week.

Micro-organisms

Pathogenic micro-organisms include bacteria, fungi and others. They kill their hosts and are relatively host-specific. Various *microbial* insect diseases occur naturally, but may also be used as biological pesticides. When naturally occuring these out breaks are density dependent in that they generally occur where insect population become denser.

Bacteria used for biological control of insect, via their digestive tracts so insect with sucking mouth parts like aphids and scales insect are difficult to control with bacterial biological control. Lepidopteran (moth butterfly), Coleopteran (beetle) and Dipteran (true flies) insect pests. The bacteria is available in sachets of dried spores which are mixed with water and sprayed onto vulnerable plants such as brassic as and fruit trues. *Bacillus thuringiensis* has also been incorporated into crops, making them resistant to these pests and thus reducing the use of pesticides.

Fungi that cause disease in insects are known an *entomopathogenic fungi* including at least fourteen species that attack aphids. *Beauveria bassiana* is used to manage a wide variety of insect pests including whiteflies, thrips aphids and weevils. A remarkable additional feature of some fungi is their effect on plant fitness.

e.g.,

1. *Beauveria bassiana* (Against whiteflies, thrips, aphids and weevils)
2. *Mycorrhizium* spp. (Against beetles, locusts and grasshoppers, *hemiptera, spider mites* and other pests)
3. *Lecanicillium* spp. (Against white flies, thrips and aphids).

Indirect Control

Pests may be controlled by biological control agents that do not prey directly upon them.

For example the Australian bush fly musca vetustissima, is a major nuisance pest in Australia, but native decomposers found in Australia are not adapted to feeding on cow dung which is where bush flies breed.

Mechanical and Physical Controls of Insect

Mechanical and physical controls kill a pest directly or make the environment unsuitable for it. Traps for rodents are examples of mechnical control. Physical controls include nulches for weed management, steam, sterilization of the soil for disease management, or barriers such as screens to keep birds or insects out.

Chemical Control of Insect

Chemical control is the use of pesticides. IPM, pesticides are used only when needed and in combination with other approaches for move effective long term control. Also pesticides are selected and applied in a way that minimizes their possible harmful to people and the environment with IPM you will use the most selective pesticide that will do the the job and be the safest for other organism and for soil and water quality use pesticides in bait station rather than sprays or spot-spray a few weeds instead of an entire area.

Two type of chemical compound:

1. **Inorganic chemical compound:** The sodium and fluorine is a powder like mixture. This compound is a soluble in water. There fore, should not be sprayed on plant. It is widely used in poison bitting mouth parts baits to control, grashopper and orthoptera insect. The more effective insecticides are available hence this insecticide is not much used now.

   e.g., Inorganic compound

   i) NaF ii) BaSiF

   iii) Na, Al, F iv) Sulphur, P, Sb, Hg etc.

2. **Organic compound chemical:** There are a wide rang of organic insecticides including vegetable oils and synthetic chemical used to control insect pests.
   i) Commercially nicotine is an alkaloid derived from tobacco. Nicotine sulphate is very toxic to insects as well as to humens.
   ii) The second insecticidal activity of pyrethrum discovered around 2000 is extracted from the flowers infloresence of *chrysanthemum coccineum.* The insecticide is commonly contained in household aerosal sprays in insect pest.
   iii) The seeds of neem (Azadirachta indica) posses four ingredient which have insecticidal properties. The suspension of powdered seed at 0.1% concentration act as repellent for desert locust but the powdered material in the ratio 5.55 serves as protectant of grains against storage pest.

D.D.T. (Dichlorodiphenyl Trichloroethane $C_{14}H_9Cl_5$): The first discovered by German Chemist Oeidler in 1874. The insecticide is a more powerful chemical. DDT is a direct contact stomach poison has long residual action. It affects the sensory organs and nervous system and causes violent agitation which is followed by paralysis and nervous system and causes violent agitation which is followed by paralysis and death. It is slow acting insecticides relatively non-toxic to mammals but in oil solution it is absorbed by the skin. It is also found to stimulate plant growth in some crops like potato, tomato etc. D.D.T. is effective against a wide range of insect pest although some of them have developed resistance to it. Since DDT kills natural enemies hence some insects such a aphids scale insect mealy bugs mite develope in more number after its application.

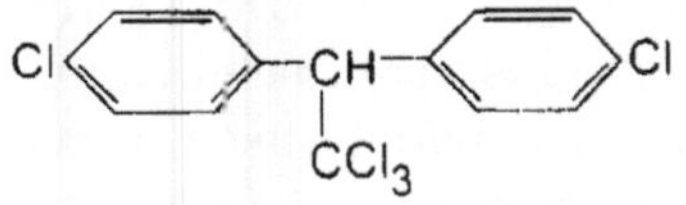

Cl CH Cl
CCl_3
$(C1_4 H_9Cl_5)$

D.D.T. Structure and Formula

B.H.C. ($C_6H_6C1_6$). Benzene hexachloride) has a wider spectrum than D.D.T. and is effective against the insect and aphids. B.H.C. is less poisonous to man and domesticated animals dog, cat, goat than D.D.T. It can be used for controlling a wide range of vegetable pests on which BHC can to be used. Some example of insecticidal control chemical.

1. Lindane $C_6H_6Cl_6$
2. Strobane $C_{10}H_{11}Cl_{17}$
3. Adrin $C_{12}H_8Cl$
4. Dieldrin $C_{12}H_8Cl_6O$
5. Endrin $C_6H_8Cl_6O$
6. Malathion ($C_6H_{19}O_6PS_2$)
7. Ethion $C_2H_{22}O_4P_2S_4$

Fumigants

The fumigant which vapourises readily at room temperature is the most useful. They may be used to control all types of insect with irrespective of type of mouth parts since the gas enters the insect body through the spiracle during respiration. The fumigants divide in some part of gas, liquid and solids.

1. Gases - CH_3B_5 Methyl Bromide
2. Liquids - $C_2H_5C_{12}$, Ethylene dichloride
3. Solids - Sodium cyanide, Aluminium phosphide

   NaCH

Toxicity of the all insect control insecticides

The toxicity of an insecticide is established by exposing test animals to a range of does and number killed at each dose. When the exact amount of insecticide being applied per body weight is know the lethal dose which kills 50% of the population can be determined (LD_{50}). If the insect have been dipped in different concentration than the lethal concentration is established LC_{50}. Once these values are calculated for different insect and for vertebrate, it is possible to compare these values to Invertebrate and Arthropody insect insecticides.

13

Important Insect Orders and Classification

The original divisions of the class insecta are a matter of dispute and it is not easy to decide which division to use many, different original order and groups are recognised by different experts and which one is best will no doubt remain problem. The words taxonomy, systematics and classification have been used alternatively for the same subject by various scientists. However Simpson 1964 has attempted to segregate these. According to him classification is the order of animals into groups on the basis of their relationship that is association of contiguity similarity or both. Systematic is the scientific study of the kinds and diversity of organism and of any and all relationships among them, while taxonomy is the theoretical study of classification including its bases, principal procedures and rule. All animal species are segregated into small and large groups on the basis of their comparative similarities and dissimilarities. The largest group is called phyllum as followed by class, order, family, genus-species and sub-species respectively. The Phylum, class, order, genus, species etc. The division of class insecta into primitive wingless insects, the apterygota and the secondary winged-insects the pterygota and included some orders. The class insecta may be divided into two subclasses:

i) Apterygota

The name Apterygota is Greek in origin and means "without wings."

These are the minute and primitive insect without wings. The insect development is generally direct and metamorphosis is absent. More than a pair of pre-genital appendages is present. Mandibles are situated on the anterior part of the head-capsule and move with the head capsule at a single point. This sub-class includes the following order thysanura under the insect silverfish and collembola under the insect spring tail insect.

This subclass included the following order (4 order only)

Thysanura	=	Silver fish
Diplura	=	Japygids
Protura	=	Proturans
Collembola	=	Spring tail

ii) Pterygota

The sub class Pterygota includes most of the world's insect species. Pterygota means "wings" and describes insect that have "wings" or once had wings in their evolutionary history. These insect also undergo matamorphosis. The subclass is divided into two groups.

1. Exopterygota (16 order)
2. Endopterygota (09 order)

i) **Exopterygota:** The insect shows incomplete metamorphosis, some time pupal instar is rarely present and wings are well developed. The insect immature stage includes Nymph. The exopterygota includes below orders. The life cycle includes just three stage egg, nymph and adult. Only the final molt stage has functional wings.

Order		Insect
1. Odonata	–	Dragonfly
2. Orthoptera	–	Grasshopper
3. Dictyoptera	–	Cockroaches
4. Isoptera	–	Termites
5. Mallophaga	–	Bird-lice
6. Siphunculata	–	Sucking insect lice
7. Hemiptera	–	Plants bugs
8. Thysanoptera	–	Thrips
9. Ephemeroptera	–	Mayfly
10. Plecoptera	–	Stonefly
11. Grylloblattodea	–	Grylloblata
12. Phasmida	–	Phasmids
13. Dermaptera	–	Earwing
14. Zoraptera	–	Zorapterans
15. Psocoptera	–	Book-lice
16. Embioptera	–	Embids.

ii) **Endopterygota:** The metamorphosis is complete and accompanied by the pupal instar. The immature stages are larvae which differ from adult and wings are developed. Endopterygota includes below orders. The life cycle includes just four stage, eggs, larva, pupa and adult. The Pupal stage is inactive.

Order		Insect
1. Neuroptera	-	Lace wings
2. Lepidoptera	-	Moth and butterfly
3. Diptera	-	Houseflies and mosquitoes
4. Hymenoptera	-	Honeybees wasp etc.
5. Coleoptera	-	Beetles
6. Mecoptera	-	Scorpionfly
7. Trichoptera	-	Caddisfly
8. Siphonoptera	-	Fleas
9. Strepsiptera	-	Stylops

Apterygota Order Characters and Examples

1. **Order: Thysanura** (Thysanos-fringed, ura-tail, bristle tails, *e.g.*, silverfishes)

Characteristics

a) They are small 0.4 – 3.0 cm in length and width in silverfishes.

b) Body elongated.

c) Mouth parts ectognathous and biting type.

d) Head Broadly sessile and very little movable, antennae long and filiform, many segmented.

e) The compound eyes large.

f) The tarsi 2-4 commonly 2 claws.

g) The abdomen 11 segmented caudle tapering behind with variable number of lateral styliform in silverfishes.

h) The tracheal system highly developed.

i) Thysanura includes two large families

i) Lepismatidea – Silverfishes *ii*) Machilidae – spring tails.

j) They are found inside houses libraries etc. as well as under tree bark etc.

2. **Order: Collembola**

   [Coll-glue, embal - a wedge; spring tails–soil insect]

   a) The mouth part are chewing types.
   b) The eyes and ocelli dorsal sides are absent, but lateral ocelli are present.
   c) Antennae short to long usually four segmented.
   d) The tarsi fused to long tabia.
   e) Abdomen six segmented in collembola insects.
   f) The furcula present on the venter of fourth segment helps in jumping *e.g.*, (spring tail).
   g) They are common in soil, leaf litter and other organic compound. They are found in generally mushroom, cereal crops, sugarcane etc.

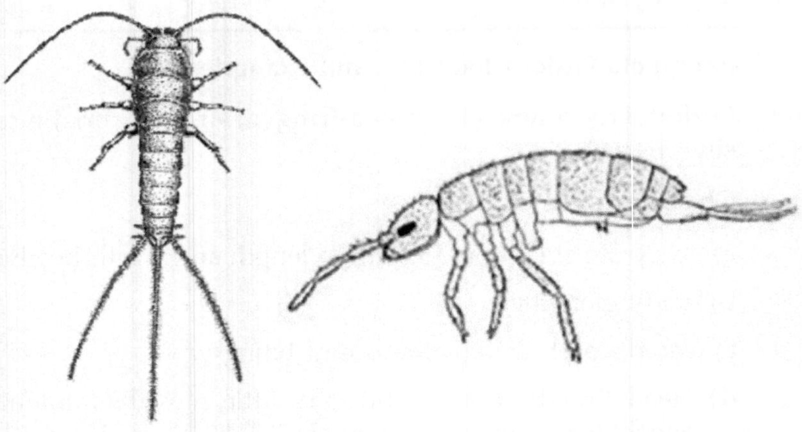

(A) Silverfish insects, (B) Springtail insect

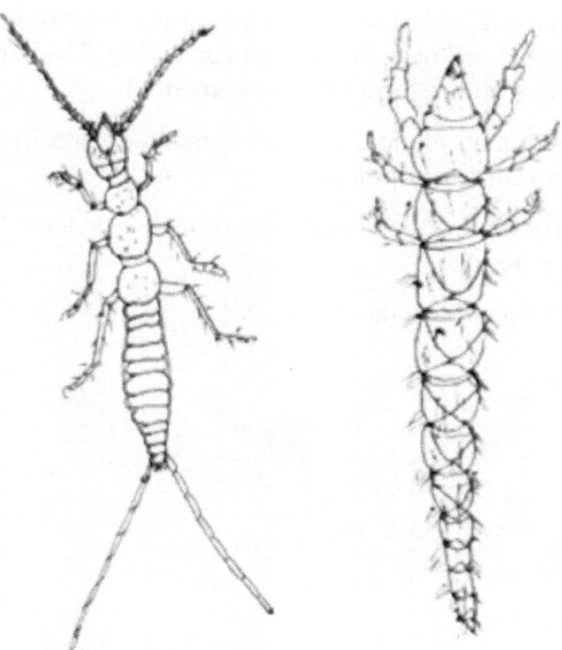

Diplura (Cambodea, Acerentulus (Acerentomon)

3. **Order: Diplura -** Dipl-two, ura-tail; two pronged bristle tail

   The minute flattened Apterygota consisting entognathous bitting mouth parts, unsegmented tarsi abdominal segments with paired lateral pregenital stylli form appendages with a pair of cerci of variable from but without compound eyes, ocelli, median terminal filament and malphighian tubules. They are found in damp situation in caves under tree bark, in the soil

4. **Order Protura:** Pro-first, ura-tail

   Minut soft bodied unsegmented apterygota having entognathous piercing mouth parts, but no eyes. Ocelli antennae and cerci is the terminal (11th) abdominal segment with a median telson. These are inhabitant of soil and leaf litter and require moist conditions. They feed on decaying organic matter. *E.g. Acerentulu* **Pterygota: Insect order**

5. **Order-Odonata Odon**: A tooth; dragonflies and damselflies.

   a) The odonata insect medium and large elongated insect.

   b) The mouth parts chewing type.

   c) The compound eyes large and 3 dorsal ocelli, antennae short and bristle-like.

d) The wing 2 pair membranous wings vanation netlike with pigmented cell near the apex of each wing. The legs are adopted for prey capture, and torsi 3 segmented.

e) The abdomen elongated, one segmented cerci in male serve as claspers during copulation

f) The larvae are aquatic with closed tracheal system and feed upon aquatic insect.

e.g., Dragonflies mosquito larvae.

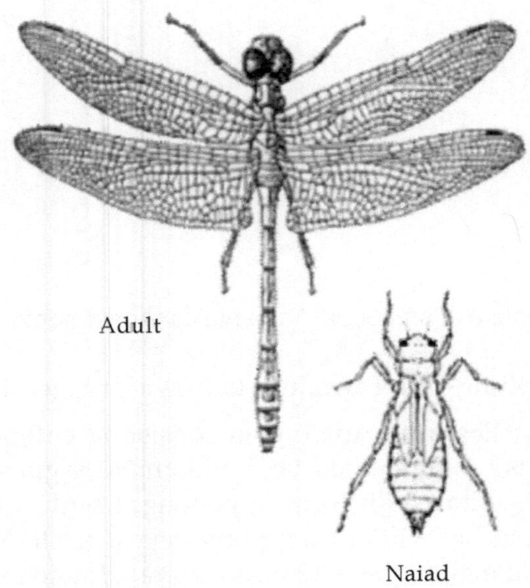

6. **Order-Orthoptera:** The orthoptera order meaning (ortho-straight, pteran = wing) and under the examples grasshoppers, locusts Crickets etc.

   a) The insect is large to small sized.

   b) The mouth parts biting and chewing type.

   c) The antennae are elongated and filiform.

   d) The insect have a well developed prothorax and two pairs of wings hind and forewings thickened and leathery in texture to form tegmina. The hind wing is membranous and folded fan like under forewings.

   e) The compound eyes and two ocelli present.

f) Meta morphosis is incomplete and the young ones look like an adult in structure and habit. All species are oviparous. Parthenogenesis very rare.

g) The tympanum is present on each side of the first abdominal segment and for tibia.

h) The hind legs usually enlarged and modified for Jumping Lege.

Orthoptera Families of Agricultural Importance.

i) **Family-Tettigoniidea** [Get their common name from their song] Characteristics

a) The long horn grasshoppers bush crickets.

b) The wingless forms, when wings present left tegmina overlap right one.

c) All tarsi are 4 segmented.

d) The ovipositor very long even exceeding body length, sword-like.

f) The insect are omnivorous and phytophagous. *E.g* Grasshoppers, crickets etc.

g) Fore-tibia possess auditory or tympanal organs.

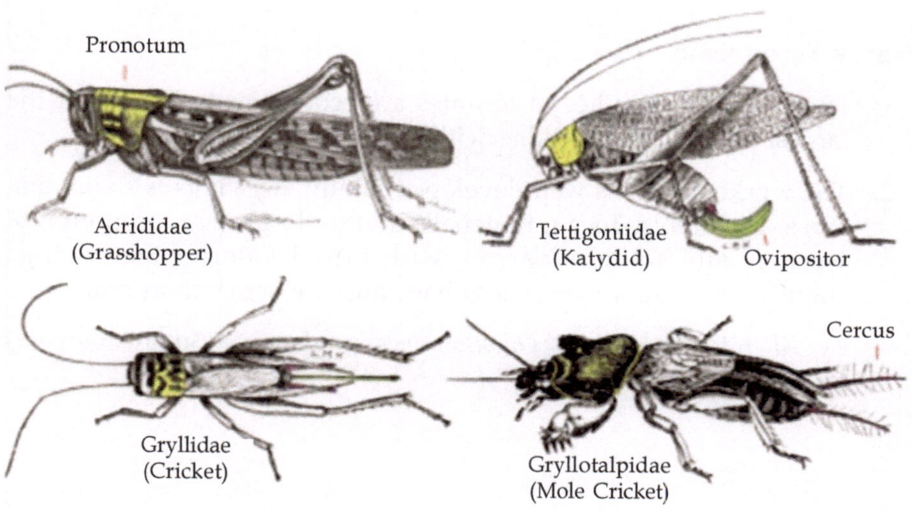

ii) **Family-Acrididae:** The insect antennae are small than body length. The prothorax extends down on the side reaching the base of fore legs but hind legs are well developed and help in the insects jumping. There is row of peg like files on the femur used for sound production. Tarsi are three segmented. The tympanum is located on each side of the first abdominal segment. The ovipositor is small, *e.g.*, Locust (Schistocerca gregaria) Rice grasshopper etc.

7. **Order: Isoptera** (Iso-equal, Ptera-wing) Characteristics:

   a) Medium sized and soft bodied insects.

   b) These are social insects and poly-morphic species living in large communities or different castes live together in colonies.

   c) Mouth part are bitting and chewing type, metamorphosis is simple, antennae smaller than the body length and moniliform type of antennae.

   d) Compound eyes and ocelli are absent.

   e) The tarsi are four segmented and cerci is small and some times rudimentary.

   f) Termites are mostly tropical live in large communites in underground nest.

   Important family of agricultural importance

Family Termididae

1. Hough all the families of termites are economically important the largest family is termididae with following characters.

2. The worker castes is well developed, hindwing wings without anal lobes, venation reduced, fontenelle and ocelli present, pronotum of workers and soldiers narrow with raised anterior lobe, wings slightly reticulate more or less hair, anterior wing short scale.

   e.g. Microtermes, beesoni, Odontotermes banglorensis (Sugarcane pest) termes, *Trinervitermes heimi etc.*

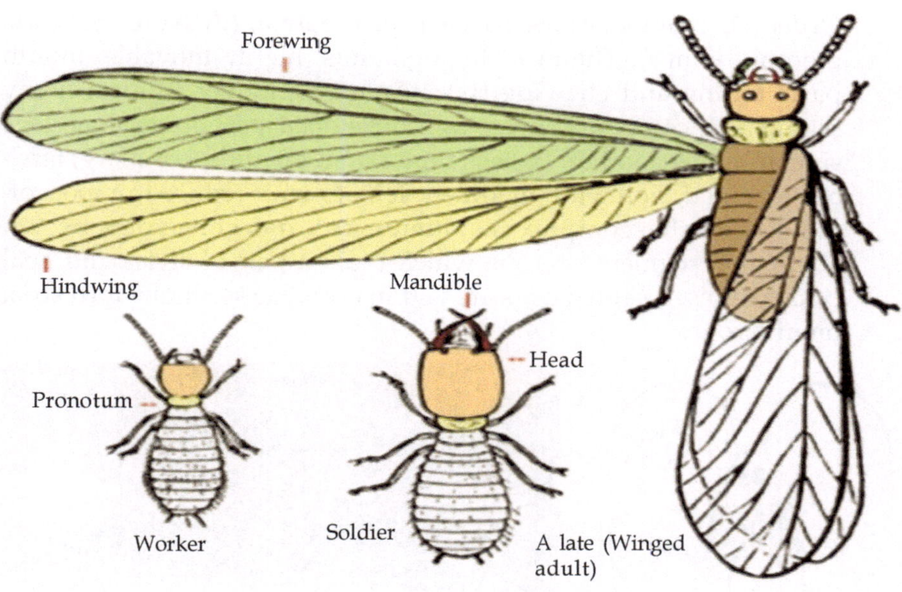

Order Isoptera (Termites)

8. **Order: Mallophage** (Mallo-wool, phaga-eat; chewing lice, birds lice)

   a) The small body dorsoventrally flattened triangular head broad than thorax, mouth parts chewing type.

   b) The compound eyes reduced ocelli absent, antennae 3-5 segmented usually filiform, wings absent, tarsi modified for grassping hairs in some species infesting mammals.

   c) The mallophago are ectoparasites of birds but considerable number of species have mammalian host.

   d) Some species feed upon blood. Few species are poultry pest.

   e.g., Common chicken louse - Menopon pallidum.

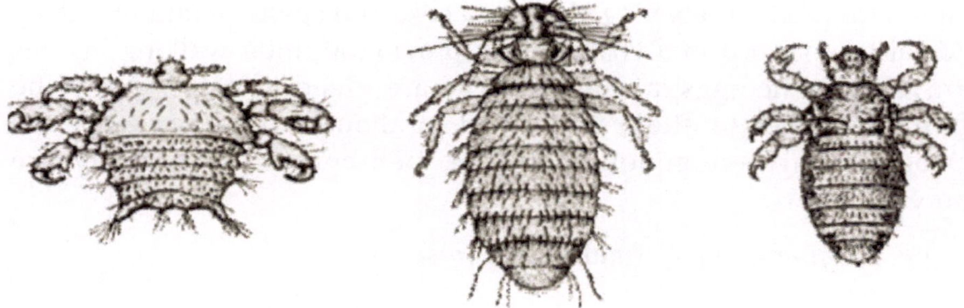

Mallophaga, biting louse, bird louse

9. **Order (Dictyoptera-Cockroaches): Dictyoptera**, (dictyon – network, pteron – wings). The head hypognathus, highly movable, mouth parts bitting and chewing type. Antennae longer filiform many segmented, compound eyes, well developed pronotum large, legs similar to one another. Forlegs raptorial in mantids. coxa larg, tarsi-5 segmented, Forewing modified, female with reduced ovipositor, male genitalia complex, asymmetrical and conceated by 9th abdominal segment, sternite which bears a pair of styles and anal cerci many segmented eggs are laid in oothecae ssentially torrestrial omnivorous.

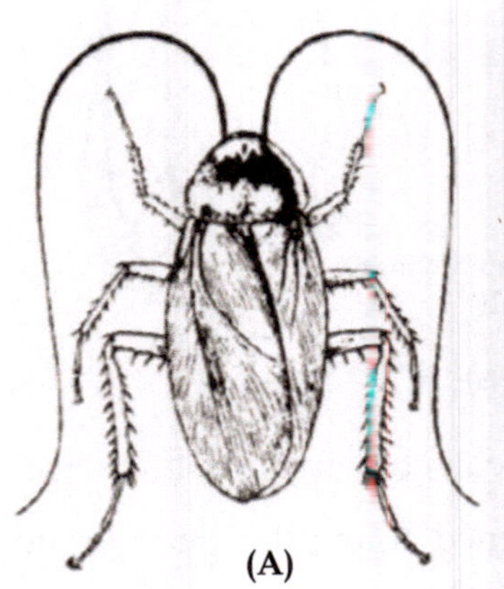

(A) Cockroaches (Periplanata-americana) (B) Preying mantide

1. Family-Blattidae: The head is almost completely covered from above by large shield like pronotum two ocelli are usually represented by fenestrae. Forlegs are not modified to raptorial type. The eggs laid in oothecae are glued to some plant. *Periplanata americana.*

2. Family-Mantidae: The prothorax elongates small triangular Head quite evident, and movable with neck ocelli 3 or absent. Fore legs raptorial adapted for predation having elongated coxa and spined femur and tibia, held raised forword in a characterstic praying attitude walking insects; carnivorous. The eggs laid in oothecae are glued to some plant. The moult 3-12 times to attain adulthood in about a year. Some species reproduce parthenogenetically. The members of this family are carnivorous.

Eg. ***Preying mantide (mantis religiosa)***

10. **Order-Hemiptera** (Hemi-one-half, Ptera-wings). The insects are medium sized, soft bodied and elongated oval insects. The mouth parts are piercing and sucking type in which labium is adapted in the form of a beak or rostrum. The compound eyes are large and antennae are four to ten segmented. Two pairs of wings are present and cerci are absent, tarsi are one to three segmented.

    The order Hemiptera is divided in two **sub-orders:**

    a) Sub-order - Homoptera - Aphids

    b) Sub-order - Heteroptera - Bugs.

Sub-order: Homoptera

Characteristics

1. Two pair of wing is present. The anterior wing pair is often of harder consistency and throughout of the same texture.
2. The pronotum is small; tarsi are 1-2-3 segmented.
3. Metamorphosis is simple exepting male scales insect and white flies the labium is attached with the head.

Iimportant families of Agricultural Importance

1. **Family-Jassidae or Cicadellidae:** The insect are generally small, elongated, cylendrical, having 8 segments and antennae setaceous type. The double rows of spines are present on the hind tibia, compound eyes are large and two ocelli are present. Thorax is narrower than head and attached with the abdomen without any prominent division. Tarsi are three segmented. Mostly rest on the lower surface of leaves and when disturbed they Leap often several feet and fly. The leaf hoppers are usually about ¼ inch in length and usually tapering posteriorly. *E.g.,*

   1. Mango leaf hopper Amritodus *(Idiocerus)*
   2. Paddy leaf hopper *(Nephotettix virescens)*
   3. Cotton jassid *(Amrasca biguttula)*

2. **Family-Fulgoridae:** Is a large group of Hemiptera insects? They are mostly of moderate to large size many with superficial resemblance to all insects. The antennae are three segmented. The head in the form of snout protruted forward, two occelli are present, the wing are membranous, small and not fully covering the abdomen,

tarsi five segmented. They are mostly grey insect and the female is provided with anal tuft. *E.g.*, Sugarcane leaf hopper (Pyrilla perpusilla).

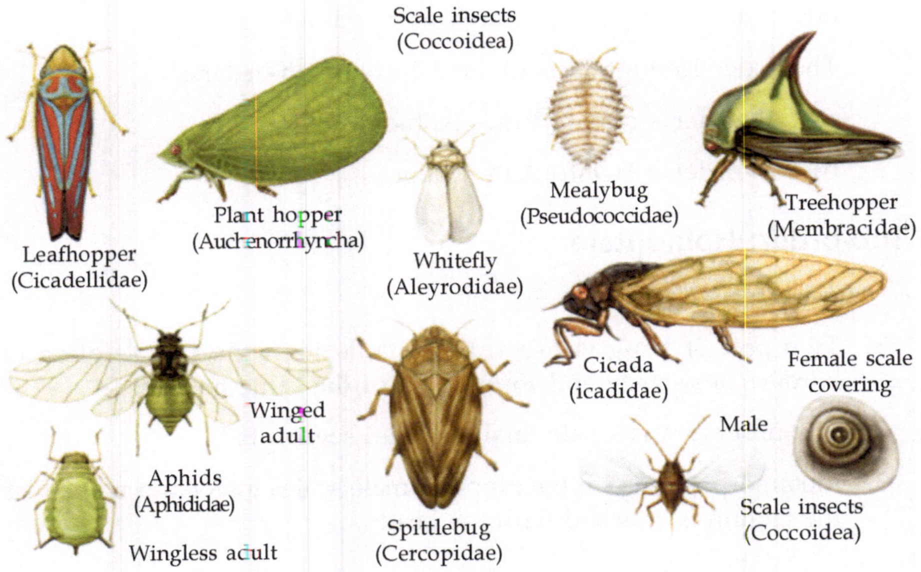

Sub order Homoptera (Family members)

3. **Family-Aphididae:** They are small and less pear shaped and wingless insects. The head is hypognathous small and thorax fused with head. The antennae are 3 or 6 segmented. The compound eyes and 3 ocelli are present. The honey-tubes are the most characteristic organs located on 5th abdominal segment, 9th abdominal tergun of abdomen anus is situated which secretes honey dew. *E.g* .Mustard aphids (Lipaphis-erysimi). Aphids are polymorphic insect.

4. **Family-Coccidae:** They are commonly known as soft scales, wax scales or tortoise scales. The insects have small body, compressed and covered with whitish meal. The female is wingless and round in structure. The male are generally winged and the forewings are well developed the legs are highly reduced and tarsi are one segmented and metamorphosis is complete in male and in- complete in female. Mouth part are absent in males. They donot feed in the adult stage. *E.g. Mango mealy bug (Drosicha-mangifera)*. In some genera they possess legs but in other they do not and the antannae may be shortened or missing.

5. **Family-Lacciferidae:** The insect are small sized with irregular globular body. The legs and abdominal spiracles are absent. The antennae are vestigial and three or four segmented. Both winged and wingless males are found having four ocelli. *E.g* Lac ***insect (Laccifer-lacca).*** All lac producing insect are scale insects.
6. **Family-Pseudococcidae:** The family is represented by the familiar mealy bugs. Their bodies are flat with waxy excretion of white powdery substance. Legs are well developed and antennae are comparatively poorly developed, *e.g., **mealy bug (Pseudococcus nialenis)***. They seldon move very slowly and feed in the same way as aphids and scale insects.

Sub-order Heteroptera

(Hetero = different, Pteron = wings). This sub-order is divided into two series.

i) *Gymnocerata:* Antennae long and conspicuous in front of the Head, under the family reduviidae, cimicidae, pyrrhocoridae, scutellaridae pentatomidae).

ii) *Cryptocerata:* The insect Antennae are short and concealed beneath in the head, of the family notonectidae, nepidae.

Some characteristics: The two pair of wing are always found in which forewing is thickened at the base and membranous. These are called hemelytra. Mouth part are piercing and sucking type. The body is usually broad and flat. The metamorphosis is incomplete.

1. **Family-Reduviidea:** Head long, antennae filiform 4-5 segmented, rostrum short, strong curved with 3 segments the tip in striated furrow between fore coxae. Some blood suckers, *E.g., Acamthaspis siva (predat or on apis indica)*. The saliva contains enzymes, that predigest the tissue and facilitates swallowing.
2. **Family-Pyrrhocoridae:** The insects are medium to large sized, antennae and rostrum 4 segmented, ocelli are absent tarsi are three segmented; prothorax is larger than the mesothorax. *E.g., Red cotton bugs (Dysdercus cingulatus).* The Brightly coloured with red/black marking and elongate oval in shape. They are insects Phytophagous.
3. **Family-Coreidae (Leaf bugs):** The body is elongated, antennae 4 segmented, eyes and ocelli well developed, rostrum 4 segmented and scent gland opens on metathorax. The Leaf like legs, tarsi 3 segmented and insect are phytophagous.The insect is dull coloured, brownish bugs. Body is oval shape. E.g., *Gundhi bug (Leptocorisa varicornis).*

4. **Family-Pentatomidae (Stink bugs):** They are medium and large sized insect bugs. Antennae five segmented and scutellum is large, triangular and never covering the entire abdomen. The insect is phytophagous or predaceous on caterpillar's compound eyes and two ocelli are present. Brightly coloured, moderate size bugs mostly greenish shield like in shape.

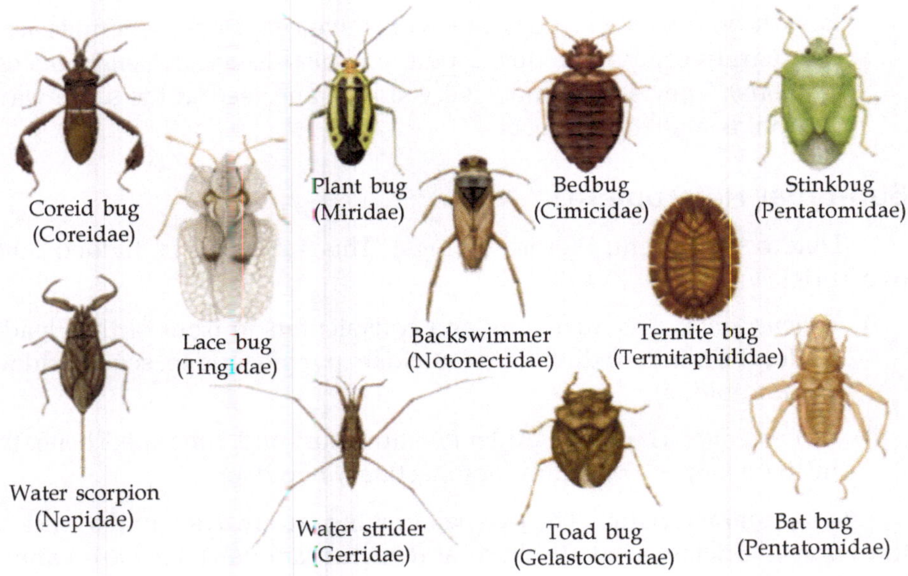

Bagrada crucifer arum (Painted bug) Nezara virdula (Feeding on the Bajara)

5. **Family-Scutelleridae (Shield backed bugs)**: The insects medium size, scutellum very large and extends upto apex of abdomen exposing the wing at its edge. Antennae short, five segmented, rostrum 4 segmented. *E.g., Chrysocoris stolli* wolf (litchi bugs, cassia occidentalies plant).

6. **Family-Cimicida:** The members of this family are ectoparasites of human beings and animals. They suck the blood. The insect body is oval and flattened, hemielytra short, rostrum small triarticulate and lies in a ventral groove. The metapleural odoriferous glands present. *E.g. , Cimex lectutarius* (Bed bugs).

B. Endopterygota under 8 orders included

Order: Lepidoptera (Moth and Butterflies)

1. (Lepidos = scales, Pteron = wings)

   Characteristics

The insect have large body covered with scales. The insect have beautiful colour and are found in the gardens, vegetables flowers etc. The head is small hypognathous type and attached with the thorax by a cervix. The eyes are compound and ocelli are generally present. Mouth parts usually long, sucking proboscis suctorial type by the maxillae. Wing two pair and female wingless covered with scales. The torsi are 5 segmented. The metamorphosis is complete and the larvae are generally called as caterpillars which are having chewing and biting type mouth parts.

This order is generally divided into two parts 1. Rhopalocera (butterflies).

2. **Heterocera** (moths): Some important Agricultural Families 1. FamilyñGelechiidae

Characteristics: Includes medium sized moth. The colour is dull. The forwings are narrower than hindwing. The maxillary palps are rudimentary. The labial palpi are three segmented. The hind tibia possesses hard hairs. *E.g.*, ***Cotton pink boll worm (pectin ophora,* gossy piella), potato tuber moth (Gnorimosc hema** *operculella*).

3. **Family-Pyralidae**: These are small sized moths with labial palpi snout like. The compound eyes and ocelli are present. Fore wing narrow but hind wing are broad. The larva bears two setae but show prolegs in a circle. The larvae feedon leaves, fruit flowers and stem. *E.g.*, **(1) Cotton leaf roller (*Haritalodes derogata*), (2) Sugar top borer (*trporyza novella*).**

4. **Family-Noctuidae**: The insects are medium to large sized and brown in colour. These are nocturnal in habit. The maxillary palpi are absent but labial are present. The antennae are filiform in female and pectinate in males, forewings narrow. The larva is of dull colours. *E.g.*, Eublemma amabilis (enemy of lac insect).

1. Plusiani (crucifer pest)
2. *Spodoptera litura* Fabr.
3. Fruit sucking moth (*Achoca janata*)

5. **Family-Cymbidae**: Includes insects of medium size, and green found among the herbage of tree and shrubs. The wings are spotted and larva have hairy covering and cocoon is boat shaped the insect live.
6. **Family-Saturoniidae:** The maxillary palps and tympanal organs absent, probascis rarely developed and antennae pectinated especially in male. These have hairy wings. The caterpillar has stout bodied with spiny tubercles. The caterpillars feed on leaves and spin large cocoons. The tibia with spines is present. E.g., Tasar silk worm (Anterea paphia), Munga silk worm (Anterea assama in green trees and cotton plants. *E.g., spotted boll worm (Earias vitella).*

Various lepidopterans species

7. **Family-Bombycidae:** The insect is medium size and yellowish in colour. Maxillary palps and tympanal organs are absent. Proboscis rarely developed antennae are pectinate in both the sexes. The larvae are elongated having anal horn on the eight abdominal segments. The larva form dense silken cocoons. *E.g., Silk worm (Bombyx mori).*

Moth	Butterfly
1. Yellowish colour, white	Bright coloured
2. Antennae filamentous	Antennae capitate
3. Two ocelli present	All ocelli are present
4. Wing horizontal	Wing vertically
5. Eggs flat and round	Eggs cigar and cylindrical
6. Larva body covered hairs	Body smooth.

8. **Order Diptera** (Di-two, ptera-wing (All flies)

The medium sized insects commonly known as all flies, mosquitoes and fruit flies. The wings are developed from mesothorax and second pair of hindwings is developed in halteres. The mouth parts piercing and sucking but sponging type mouth parts under mandibles are generally absent. The tarsi are five segmented and metamorphosis is complete. The larva generally called as maggots which are legless. The insects omnivorous, carnivorous or sangivorous (only female), some transmit human diseases like malaria disease, sleeping sickness kalaazar etc.

Diptera order is divided into 3 sub-orders

1. **Nematocera -** The antennae is many segmented usually longer than head and thorax, all segments are alike, arista absent.
2. **Brachycera -** Antennae three segmented, style like, labial palpi 1-2 segmented. The insect head and pupa, larva obtect but usually is retractile with vertical bitting mandible.
3. **Cyclorrhapha -** The antennae 3-4 segmented, labial palpi 1 segmented, larva with vestigial head etc.

Agriculturally Important Families

1. **Family-Psychodidae**: The body and wings are clothed with coarse hair and scales present but ocelli absent. The antennae 10-14 segmented, four branched veins, larvae usually aquatic. *E.g., Trypanosoma gambiense* cause African sleeping sickness in man.

2. **Culicidae-Family**: The mouth parts are piercing slender and elongated. Ocelli absent, Antennae 10-14 segmented, pedical large plumose in male and pilose in female, hind legs raised while sitting larvae metapneustic with large thorax. *E.g., Mosquitoes* cause malaria, Filaria, yellow fever etc.

3. **Family-Tabanidae**: In cludes medium sized insects and bristleless flies. The antennae is 3 segmented annulated but devoid of styles. The eyes are large, ocelli absent. Mouth part is piercing type. Females are blood suckers. The leg pulvilli and arolium are large, and padlike. *E.g., Horse flies (Tabanus* rubidus)

4. **Family-Drosophilidae**: Includes small insects and chubby flies with large light eyes, the life-cycle is completed within a week. The forewings are present and hindwing develop the halters (Blancing parts). The antennae are jointed and rounded. The arista is dorsal and pectinate. ***E.g., Drosophila melanogaster*** (Fruit flies).

5. **Family-Muscidae**: Includes small and large fly insects. The veins are not reaching the apex of wing, lower calypter longer than prosternum. The antennae are segmented. Forewing is present and Hindwings modified into halters. The insects are present in houses. *E.g., House flies (Musca domestica, Glossina palpalis.*

6. **Family-Tachinidae**: Includes the small and parasitic insect. The head is large and movable type. The scutellum is not developed. The antennae are hairy and naked. The family members attack larva of lepidopterous. E.g., *Sturmiopsis inferens* (The silkworm parasite).

7. **Family-Agromyzidae**: Includes small to medium flies, antennae are short. The larvae are usually leaf like. The insect are winged species and tarsi different, femur is without spine. Post-vertical bristles are divergent, anal cell is present. E.g., Pea leaf miner (Phytomyza atricornis).

8. **Order Hymenoptera** (Hymen-membranous, ptera-wing): The insects have two pair of membranous wings, venation reduced and hindwing small, forewing hooklets types. Mouth parts is biting and sucking type. The eyes are compound and well developed ocelli 3-present. Ovipositor is modified into stinging apperatus. Tarsi are

5 segmented and provided with a pair of claws and empodium. Metamorphosis is complete. The larvae are grub like, chewing mouth parts. Parthenogenes is very common, male are usually halploid. The Hymenoptera are generally recognised as the most advanced group of insect.

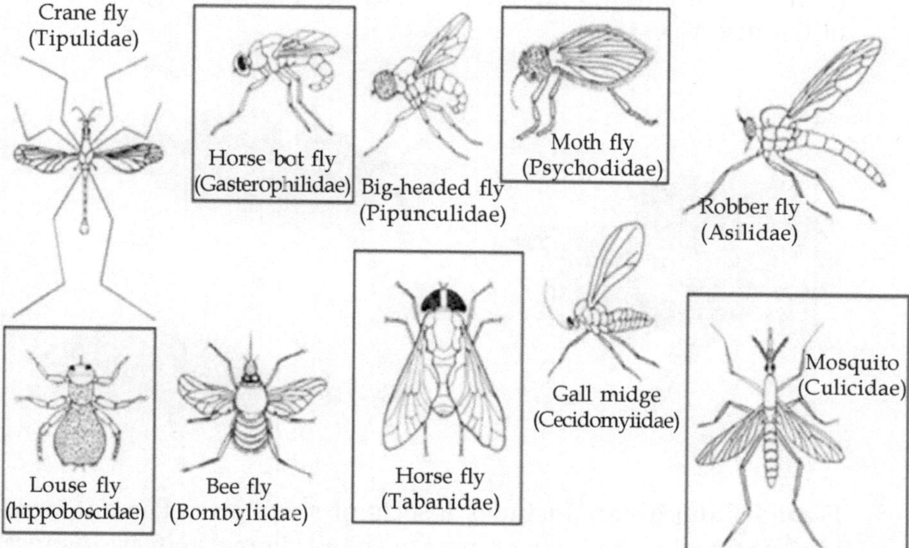

Important Agricultural Families

1. **Family-Bombidae**: The insects are black-red and yellow. The body is covered with densely hairs, antennae geniculate, glossa long, eyes not reaching base of antennae, are good pollinator insect. *E.g.*, Bumble bees.

2. **Family-Aphidae**: The insect includes a vast range of solitary and parasitic bees and social species. The head is broad as thorax. The pronotum end is small round lobe. The body is rounded and hair present in body. The maxillary palpi are one segmented and labial palps 4 segmented. The honey bee's colony composed of three castes. (*i*) Queen, (*ii*) Workers and (*iii*) Males (drones). Show pollen collection, mostly social, pollinate flowers of plant of economic importance commercial. ***E.g., Apis Indica, Apis dorsata, Apis florea, Apis mellifera etc.***

3. **Family-Tenth redinidae**: Includes medium sized insects. The posterior region of scutellum is separated as postscutellum. The ovipositor is saw like. The larvae are caterpillar like. Bear thorax leg. Antennae 9 segmented, third segment short. ***E.g., Mustard sawfly (Athalia proxima).***

4. **Family-Braconidae:** Includes small sized insect and parasitic ones. The compound eyes are naked and 3 ocelli are present. The hind femur is thick and small, divided trochanters, forwing with pterostigma at distal end of costa and costal cells are narrow. The larvae are pripneustic into several species. *E.g., Diaeretiella rapa* (Parasitic on brassica Aphids). Mostly is used inbiological control of the insect pest.

5. **Family-Eulophidae**: Includes insects of small size. The antennae are 2 segmented. Forewing is narrower and stigmal vein are distinct. The tarsi 4 segmented**. *E.g., Tetrastichus pyrillae* (Parasitic on eggs *of pyrilla perpusilla*).**

6. **Family-Trichogrammatidae**: Includes medium sized and minute species. The forewing are broader, tarsi 3 segmented and antennae with one ring-joints, axillae extended back to tegulae, parasitic on eggs of lepidoptera etc. *E.g., Trichogramma* ***minutum (widely used in biological control of insect pest).***

7. **Family-Formicidae**: These are social insect and colony consists of different species namely, workers males and females. The antennae are geniculate type. Submentum and mentum separate eyes and ocelli present in male but vestigial in female, wings present in sexual forms. The joints of head, thorax and abdomen are very clear. Labrum is vestigeal, maxillae 1 to 6 segmented; when the first larva appear, they are fed with a special nutritive secretion of the salivary gland and as soon as the first workers appear they go forth into the world Foraging (*E.g., Ants*). They take overall duties nest building, cleaning, nursing and fighting etc.

8. **Order Coleoptera:** The insects are medium and large size. The head is hypognathous, flat, cylindrical into a snout. The eyes are

compound and large, ocelli absent. The body is hard with forewings modified into horny or leathery elytra and hind wings membranous and protected by front pairs. The mouth part are biting and chewing type, metamorphosis is complete. The antennae are 11 segmented and setaceous type. Prothorax large and mobile but mesothorax reduced, legs typically cursorial some adapted for swimming, jumping. The larvae of beetles are known as grubs. While the snout beetle (weevil) grubs are legless. Coleoptera forms the largest order in the animal kingdom having about 34,000 species. The order includes two major suborders. (*i*) *Adephaga,* (*ii*) *Polyphaga* which can be identified by the following comments:

Adephaga: The wings usually show two cross-veins forming an oblongum, notopleural sulcus present in prothorax, testes tubular, and malpighain tubules four and simple. Hind leg coxae are immovable and fixed.

Polyphaga: The hind coxa movable and wing veinated, testis folicular, prothorax distinct, malphighian tubules different type.

SOME IMPORTANT FAMILIES

1. **Family-Coccinellidae**: They are small, round, convex and coloured full beetles. The insect is red to black and some spots are present. The antennae are small and clavate, Tarsi arise at 4th segment and third segment bears pad, second and fourth bearing the claws. The wings are developed, hard covered abdomen. The insects are carnivorous or phytophagous some species are predaceous on aphids and scale insects, while others are plant feeders. *E.g., Coccinella septempunctata* (Coloured full Beetle).
2. **Family-Dermestidae**: Includes insects of small size and oval beetles. The head is hypognathous and antennae are small with 11 segments. The last segment is clavate type. The compound eyes are large and ocelli present. The wings are simple and cover the abdomen. The larva is caraboid and remain covered with dense hairs like household pest. ***E.g., Trogoderna, granarium (Khapra beetle).***
3. **Family-Bruchidae**: The insect is small insects and short beetles. The antennae are 11 segmented; the last segment is clavate type. The head is prolonged into short quadrate beak. The wings arc type and tip abdomen is exposed. The Tarsi are 5 segmented. The first instar larvae has legs, antennae and eyes well developed and IInd instar larvae is devoid. Serious pests on stored pulses. ***E.g., Callosobruchus chinensis (Pulse-beetle).***

4. **Family Chrysomelidae**: The insect is small to moderate sized. The antennae are large eleven segmented and filiform. The tarsi four segmented and third segment is kidney shaped structure. The wings are simple. The abdomen is short and five segmented. The grub is stout with short legs and Antennae. The insect is leaf like and flat. *E.g.,* ***Raphido palpa foveicollis*** **(Red pumpkinbeetle).**

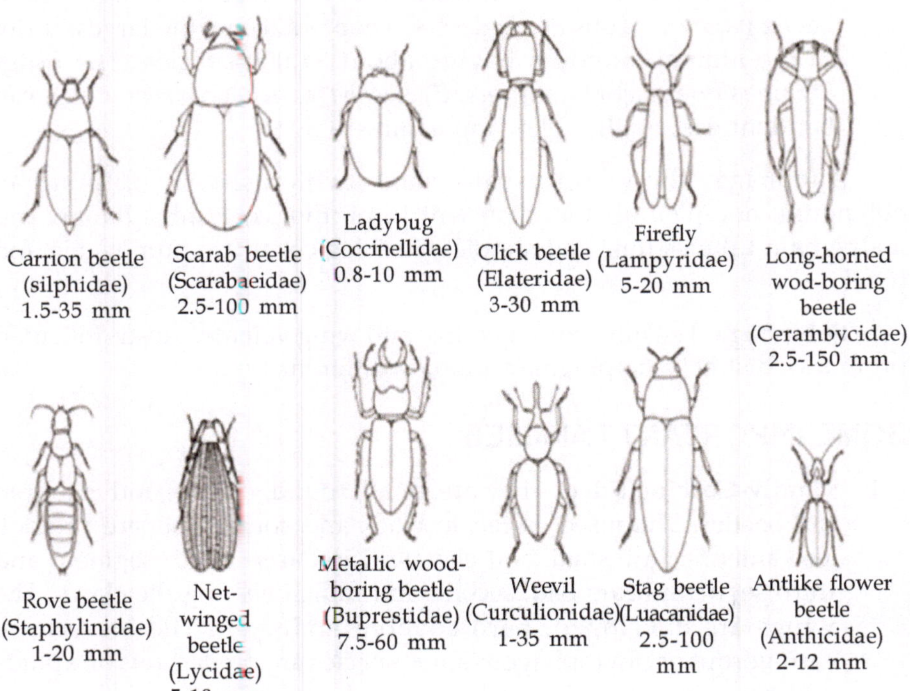

5. **Family-Curculionidae**: Head more or less produced into a rostrum, the antennae geniculate and clubbed. The labrum absent, trochanter very elongated, the body, elytra are less and very hard. Tarsi are five segmented. First and second segment are fused in Abdomen. The larvae are legless grubs, generally having dark head and white body. ***E.g., Sitophilus oryzae*** **(rice weevil).**

6. **Family-Tenebrionide**: The body is hard often globular type. The antennae are short under the projection of the side of head. The elytra fits closely to abdomen, some species have no wings. Mostly scavangers and few feed on stored grains pests. The larvae are elongate and cylindrical in which the hind end, bears two dorsal hooks. ***E.g., Tribolium castaneum*** **(Red flour beetle).**

14

Insecticide Application Equipment

Efficiency of pest control by insecticides is dependent on the correct type of pesticide formulation applied at the proper time and correct dosage, besides the good equipment. Equipment used to apply insecticides must be suited to the crop (or other situation in which the pest is to be controlled), to the scale of operation, and to the formulation of chemical to be applied. In the long run, it is good economy to buy the best equipment available.

Application equipment of high quality means not only good, durable construction but also good design that gives it a professional appearance. Because it is desirable and often necessary to utilize pesticides in a variety of ways, various types of dispensing equipment have been developed. This equipment is placed in the following categories: (1) Sprayers, (2) Dusters, and (3) Granule Applicators. In addition, specialised equipment of various sorts (such as mist machines and fog generators) are available for particular purposes. Since detailed consideration of pesticide application equipment is out of the scope of this book, only a brief consideration is given here to some important features of the main types of equipment.

1. Sprayers

Sprayers vary greatly in size from the hand-pumped flit pump with a tank capacity of as little as 1 cup to large hydraulic machines powered by gasoline engines and having tanks which can hold several hundred litres of pesticide formulation. All sprayers must provide some means of breaking up the spray liquid into droplets *(atomising)* and of conveying the droplets onto the object to be sprayed. There is always a tank, a method of pressurizing the liquid, a delivery line leading to a valve, and another delivery line leading from the valve to a nozzle. In smaller machines hydraulic pressure provides the means both of atomising the spray liquid and of imparting momentum to the droplets so that they carry to and impinge on the target. Many larger machines, however, especially those designed for low volume application; employ a fan generated stream of air to carry the droplets. They are referred or as air assisted or air blast sprayers. Automization in such sprayers may be through normal hydraulic nozzles or by power operated spinning discs or cages, which provide more mix form droplet size than conventional nozzles.

i) **Compressed air sprayers:** A compressed air sprayer is a small, hand-operated, hand- carried spray equipment with a capacity in the range of 6 to 15 litres. Air pressure is supplied by a hand-operated pump and is contained in the tank above the surface of the liquid to be dispensed. This air, usually compressed to 20-50 pounds pressure, forces the pesticide out through a discharge tube to the nozzle when the discharge valve is opened. The liquid is not mixed with air but is rushed out as a wet spray without automization. Some of these sprayers are equipped with a pressure indicating gauge. Compressed air sprayers are used for insects in lawns, orchards, home and in poultry houses. They are used by public health workers for applying residual deposits of insecticides in homes and other buildings to control mosquitoes and other insects of medical importance. Hand-pumped are compressed air sprayer. Large arrow shows of air presence in liquid and small arrows show direction of liquid flow.

ii) **Knapsack sprayers:** A knapsack sprayer is carried on the back of the operator and the pressure is created by the user working a lever up and down. It has a tank capacity of 8 to 20 litres and has higher pressures generated than a compressed air sprayer, so it is better suited for large gardens.

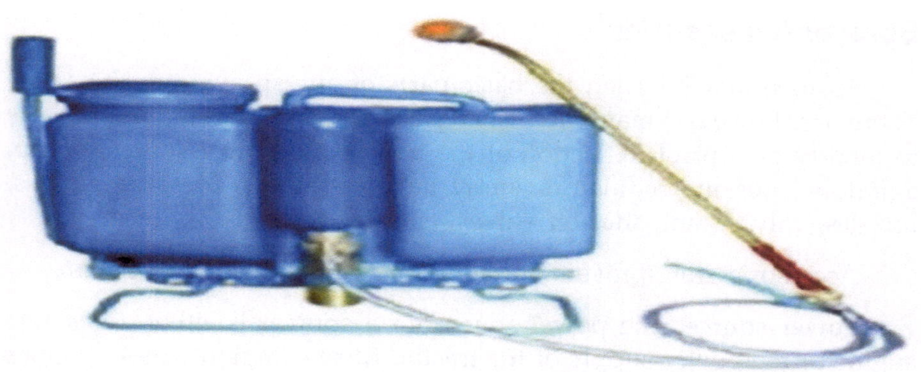

Knapsack Sprayer, Bucket-type Sprayers

iii) **Hand-actuated hydraulic pumps:** Such pumps are used with sprayers of small capacity. In this type of sprayer the user squeezes a trigger to activate a plunger this plunger displaces the liquid inside the pump, forcing it out the discharge nozzle. When the plunger is released, it causes the pump to fill up again with liquid. Such a sprayer produces a fine jet of pesticide or a fine mist, depending on the nozzle being used. These small sprayers are used for localized applications where large volumes are not required, such as in cockroach control, treating furniture, etc.

iv) **ULV devices previously:** All sprayers were of the hydraulic type and applied pesticides in very dilute form, almost wetting completely the target surfaces. But the development of air blast machines has permitted the use of chemicals in more concentrated form. In this *low volume is spraying* the spray is in very fine droplets which do not give complete cover of the target. The larger machines are suitable for use outside for mosquito control as well as for use inside in large areas. ULV sprayers require less frequent filling of the spray tank compared to high volume spraying. This is very important in aerial spraying and can materially reduce application costs.

ULV fogger, ULV sprayer, cold

Sprayer Accessories

As mentioned earlier, the basic parts of any sprayer are the same. There are, however, many accessory parts which are added to sprayers to modify or control the action of the basic parts power source, pump, agitators, pressure regulators, spray tanks, strainers, pressure gauges, nozzles, valves, tank shut-off valves, and hose.

Accessories for sprayers are addition to the basic components.

Power source: The power source for a sprayer is either a gasoline engine or the power supply of the tractor. Most commonly used engines develop 1 to 3 horse powers.

Pump: It is one of the most important parts of the sprayer. The piston and plunger type of pumps develop pressure by pushing directly on the liquid in a cylinder.

Agitators: When using power-driven equipment it is frequently necessary to maintain an active agitation of the liquid in the tank in order to keep it thoroughly and evenly mixed. If emulsion type materials are being used, a simple return by-pass as found on most power sprayers is sufficient to keep the emulsion in suspension. Mechanical agitators provide excellent mixing of the chemical for wettable powder solutions. Just be sure that your pump has adequate capacity to supply both the agitator and the nozzle.

Pressure regulators: In all power pump operated equipment it is necessary to have a pressure regulator located in the line between the pump and the discharge valve to the nozzle. The pressure regulator is set at the desired pressure for spraying and, when the pressure in the line exceeds this pressure, the regulator opens This permits some of the chemical to return to the tank through a by-pass line thus lowering the pressure at this point.

Spray tanks: Spray tanks ean be made of wood, steel, stainless steel, glass, fibreglass, and plastic. However, tanks should be strong and durable. The most suitable tanks for use in everyday pest control operations are stainless steel and fibreglass.

Strainers: A strainer is necessary in all sprayers to prevent damage to pumps and to prevent clogging of nozzles. In large equipment a strainer is placed between the tank and the pump to prevent coarse particles from reaching the pump and damaging it. Also, another strainer is frequently placed in the nozzle itself to hold back any particles liable to block the nozzle tip.

Pressure gauges: To prevent putting too much pressure into the tank as well as to set the pressure regulator, the pressure gauge is absolutely necessary. This gives an accurate representation of the pressure in the tank. Also, it is important to install a pop off valve. This device is preset to the maximum pressure which is to be put into the tank and will open and release the pressure automatically if this pressure is exceeded.

Nozzles: An important aspect of spraying is adequate coverage of the target. The primary function of a nozzle is to obtain uniform distribution and particle size of pesticides whether the material is placed on a surface or is dispersed into the air. Common patterns used in different spray nozzles are the fan, pin stream, solid cone, and hollow-cone. There are adjustable nozzles which permit the user to change the form of the spray pattern to suit the conditions under which he is spraying. Majority of spray nozzle tips are made of brass, stainless steel or carbide.

Valves: Valves provide a means of turning on and shutting off the flow of material to the nozzle in a continuous flow sprayer. Some valves are turned; some are squeezed while some are bent to cause them to operate.

Tank shut-off valves: Such valves should be provided between the tank and the hose, so pressure can be shut off from the hose without shutting down the power source driving the equipment.

Hose: The hose on any sprayer should be long enough for the purpose intended, should be of sufficient diameter to carry an adequate flow of chemicals and should be made of materials which will not be deteriorated by the pesticides and solvent being used. Different materials like neoprene (synthetic rubber), polyethylene nylon and tygon (plastics) are ased in making the hoses. Where high pressure spraying is to be done it is necessary to use heavy duty hose which will withstand high pressure.

2. Dusters

Since dusts are used far less than sprays as a means of applying insecticides to plants, there has been less development of dusting equipment. Dusters, like sprayers, may be small or large, hand or power operated. Most dusters incorporate a metering device of some sort to regulate dust output and a blower to carry the dust to the target. Large, power operated dusters are best suited for agricultural work and some types of mosquito control. There are also available small compressed air tanks for applying dust. Air is compressed in the tanks and when dust is desired the working of a single lever causes the compressed air to discharge dust make a drawing of a duster in your notebook.

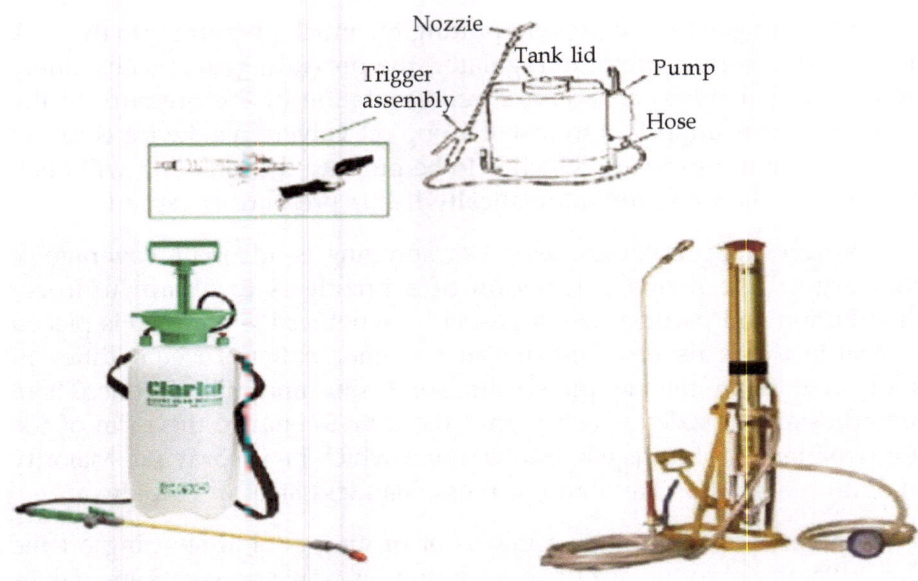

Stewardship, Garden Sprayers

Foot Sprayers

3. Granule Applicators

Various types of granule application equipment are available. Selection for a particular purpose depends on whether the granules are to be spread (*e.g.*, over pasture) or applied in a restricted manner. Most granule applicators incorporate a mechanism to feed granules in regulated amounts from the hopper. For the in-furrow application, granules may be simply released "down the spout" into the seed furrow. The granules so put are lightly "incorporated" into the soil by incorporation wheels, chain drags or press wheels. For the broadcast application there is usually some means of throwing the granules to form a swath, e.g., rapidly rotating ridged disc or oscillating arm. Granule applicators, like sprayers, are calibrated. The amount of granules spread by applicators depends on the size of the feeder-gate setting (from which granules flow), speed of travel, field roughness, and the flow rate of granules (which is affected by size, shape, texture, and density of granules, as well as by temperature and humidity.

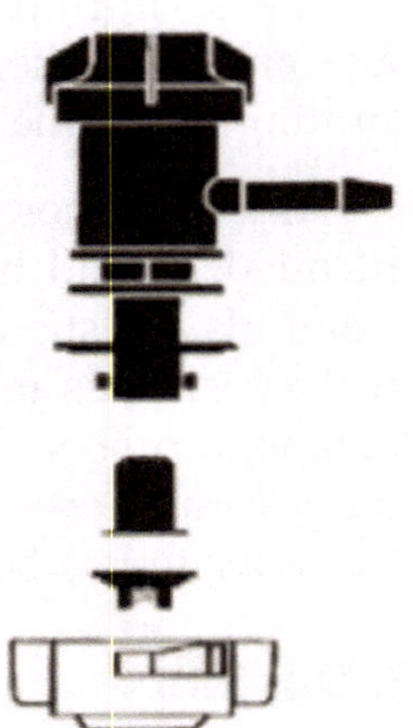

Nozzle assembly

15

The Post Embryonic Development of Insects

The insect larva after the complete development emerges out from the egg by rupturing the egg shell chorion. It passes through several stages called as instars. Each instar is separated by a moult; the growth from larva to adult involves some degree of morphological changes or metamorphosis. The egg hatches into 1st instar larva which later moults to the second instar larva and so on until at a final moult the adult imago stage. The insect moult takes place after attaining adult except in the insect apterygote insect. In family pantatomide full insect moulting occurs. Many insects especially those with a small number of instars (mosquitoes) and (*chrysocoris stolli*) insect moulting occurs as Ist, 2nd, 3rd, 4th and 5th imago instar and then after sometime adults insect emerges.

Metamorphosis of Insect: The development process by which first instar immature stage is transformed into the adult is called metamorphosis, which means literally change in the form. The developing stage is morphologically different from the adult. The degree of difference varies from slight to extreme with many intermediates stage. ***E.g., Cockroaches, Butterfly, Chrysocoris stolli Wolf***. The insect may be grouped on dependance of degree of metamorphosis as:

Types of Metamorphosis

1. **Ametamorphosis of insect:** The primitive insects, wing less or apterygota only show slight changes, brought about during post-embryonic development at each moult and moulting continues even after the final stage is reached, this type of metamorphosis is found in insects belonging to the protura thysanura, collembola of apterygota groups. Silverfish, Butterflies and moths Grasshopper etc.
2. **Paurometamorphosis:** The immature stage is called as nymph in whom the wings and genital organs are not developed. This type

of metamorphosis is met in most of the exopterygota insect like termites order Isoptera, orthoptera grasshoppers and Hemiptera order the bugs known as paurometa bolaus.

3. **Hemimetabola metamorphosis:** This is also an incomplete type of metamorphosis in which accessory organs known as gill are found in nymphal stage. The image immature stage is called naids. The insects belonging are order odonata dragon flies, etc.

4. **Holometabola metamorphosis insects:** This type of metamorphosis is also known as complete metamorphosis, take place in endopte rygotae insect in which imago immature stage is larvae stage. The larva and pupa are quite different from the adult and morphological larva forms.

Types of Larvae of Insect: The immature stage of insect species has a wide variety of forms. There are different name for the larvae of certain group of insect, dragonflies, termites, bugs, larvae of flies beetles, wasps and honey bees etc.

1. **Polypods larvae:** The larvae have well developed head, short antennae and short thoracic and abdominal legs. The caterpillar-like larvae has cylindrical body. *E.g., Moths, butterflies.*

2. **Protopod larvae:** The larvae are found insome parasitic form of order Hymenoptera. The larvae represent a very early stage of development in which little segmentation of the body has occured and cephalic and thoracic appendage is either absent. *E.g., Componatus solanopus, Ants.*

3. **Apodous:** The larvae are legless Robust or Spindle Shaped. They are divided into three types of larvae.

   a) *Eucephalous larvae in wasp:* They have more or less well developed head capsule with relatively less reduction in cephalic appendages. *E.g., Larvae of mosquitose.*

   b) *Hemicephalous:* The head and appendages are reduced in larvae. They are retracted thorax. *E.g., Housefly orders diptera.*

   c) *Apoid:* These are robust with well developed head, cared for by daily feeding in provisioned blood cells. *E.g. Ants and bees,* etc.

4. **Oligopods:** The larvae are characterised by the absence of abdominal prolegs, thoracic legs are well-developed. They may be further divided into two groups.

a) *Carabiform:* The larvae resemble the larvae of carabid beetles which are similar to the campodeiform type but have shorter legs and cerci. *E.g., Order: Coleopter, All- beetle.*

b) *Platyform:* The larvae have a flat body and short thoracic lege. *E.g., Order: Lepidoptera – Butterfly.*

c) *Scarabaeidform:* They larvae have C - Shaped body. They have short thoracic Legs, body without abdominal legs. *E.g., Order: Coleoptera.*

Types of Pupae: All endopterygote insect pass through pupal stage during development in between the last larval stage and adult stage. This instar is different from the larvae as it is more or less quescent and does not feed. These pupa are of following type.

i) *Pupa-Decticous Type:* This type of pupa is provided with functional mandibles. The appendage is always free. It is found in lepidoptera order.

ii) *Pupa-Adecticous Type:* This type of pupa is found in other endopterygote and has no functional mandibles. They are divided in two types. *E.g., Coleoptera, Hymenoptera, Diptera* etc.

1. **Hypermetabola:** This is more complex type of metamorphosis in which there are several type of larvae different from each other after each instar.

   E.g., Ist instar active,

   IInd instar robust and sluggish,

   IIIrd Instar apodoess.

   All present in order coleoptera, Diptera, Neuroptera etc.

2. **Polymetabola:** In some insect from a single egg more than one larva have been seen to hatch out and this process is termed as poly embryony and metamorphosis is known as poly metabola. Example member of families like insects etc.

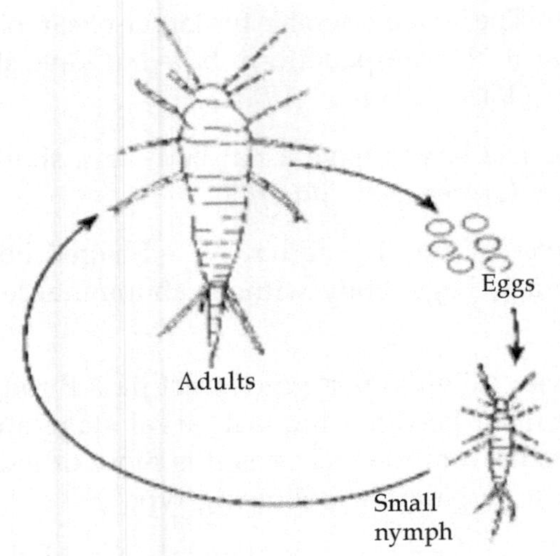

Metamorphosis in silverfish

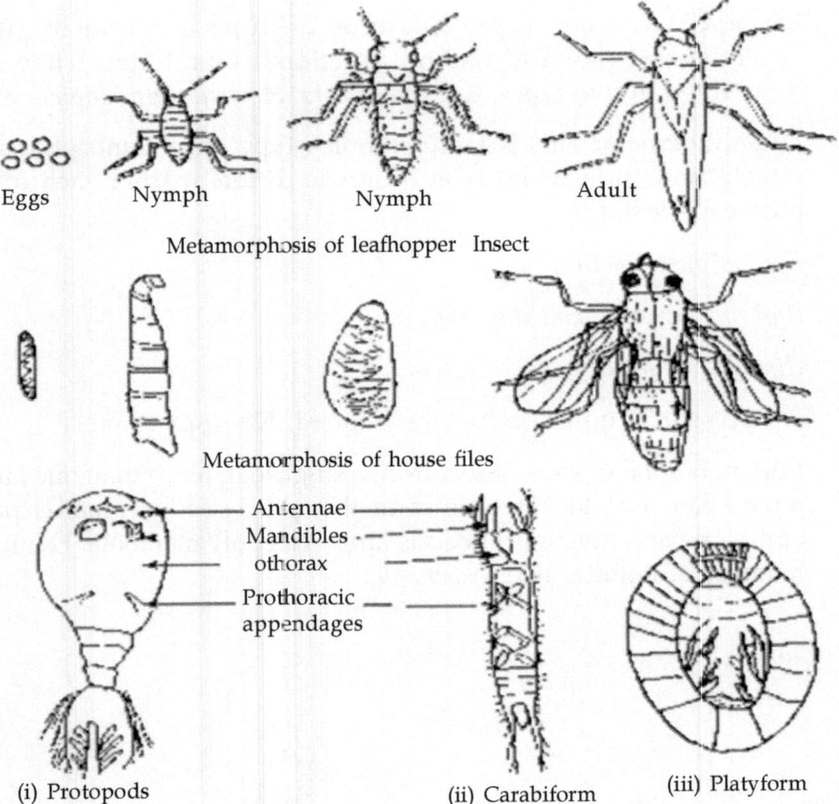

Metamorphosis of leafhopper Insect

Metamorphosis of house files

Types of larvae

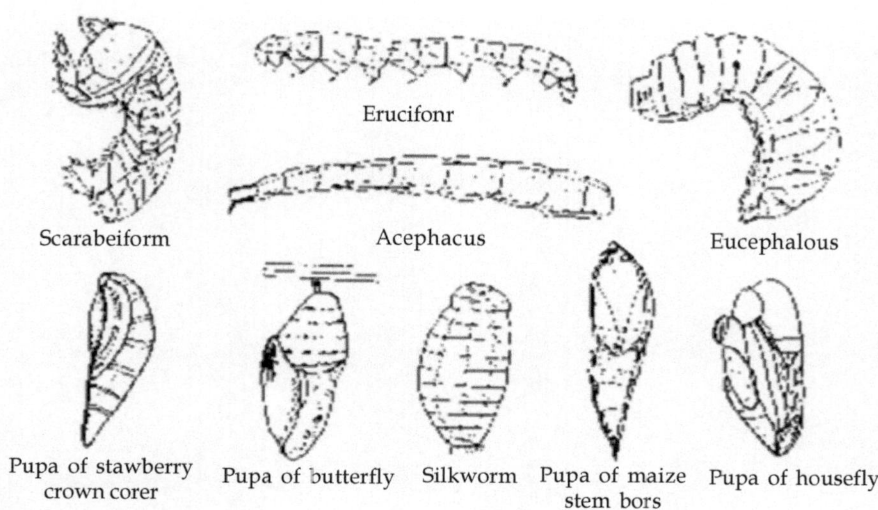

Types of pupa

16

Locusts: Life History and Control

Locusts are injurious to man as it destroys several human commodities. Locusts are phytophagous and feed on several plants cultivated by man. The locusts are insects belonging to the order orthoptera. The most popular insect is grasshopper. The only difference between the two types of insects is that locust can exist in two different behavioural states whereas grasshoppers do not. When the population density is low locusts behave as individuals, much like grasshopper. As the population density is high locusts gregariously behave as band of nymphs or swarms of adults. The changes in body shape and colour and in fertility, survial and migratory behaviour are accompanied by phase change. The most important locust species is the desert locust (Schistocerca-gregaria). The international pest influences various parts of Asia, Africa and Europe. Other locust species are highly injurious to mainkind like Bombay locust and Indian migratory locust (Locusta-migratoria) found all over India in its solitary phase. The monsoon months in July and August suitable are for locust multiplication conditions. The new generation produced migrate eastward into the Indian desert areas in May and June.

Life-cycle of Locust

The female lays eggs in eggpods containing 45-120 eggs at a depth of 7-15 cm in moist sandy soil. The female may oviposit thrice at weekly interval and lay about 500 eggs in her life-time. The eggs hatch into small nymph within 2 weeks of incubation period. The nymph has a tendency to congregate in groups and begin to march from place to place eating up all vegetation along their way. There are five nymphal instars in insects. The metamorphosis complete in the insect in 4-6 weeks.

Control

The locusts can be controlled in all its stages of life cycle. Egg, nymph and adults. By digging, the eggs could be destroyed either by ploughing or harrowing the fields. The hopper and adults may be controlled by adopting following methods.

1. **Trenching:** Trenches of about 45 cm deep and 30 cm wide may be dug at some distance in front of marching hopper bands. The trench is dug to bury the live driven hoppers.

2. **Poison Biting:** Poison baits consisting of wheat or rice bran, an insecticide and an attractant like molasses is spread around the bushes. Hoppers rest at night and molasses is spread at rate of 20-30 kg/ha during day-time.

3. **Spraying of insecticides:** Although in past BHC was dusted and aldrin emulsion was sprayed over the hoppers to kill them but now-a-days both the insecticides are banned by the government of India. Some insecticides are :

   a) Dieldrin is the epoxy of aldrin and is one of the most persistent chamical. It is used in the situation where long laiting residual effect is advantageous.

   b) Malthion ($C_{10,}$ $H_{19}O_6$ PS_2): it has been used for all types of agricultural insect pests and household insects.

   c) Methyl: It is less stable and is toxic. Its use in India is permitted only on those crops where honeybees are not acting as pollinators.

   d) Diazinon: The most common household and garden spray. It is 25% wettable powder and 25% emulsive concentrate.

   e) Phenyl pyrozole: It is very toxic to human and this compoud is more stable than aliphatic ones.

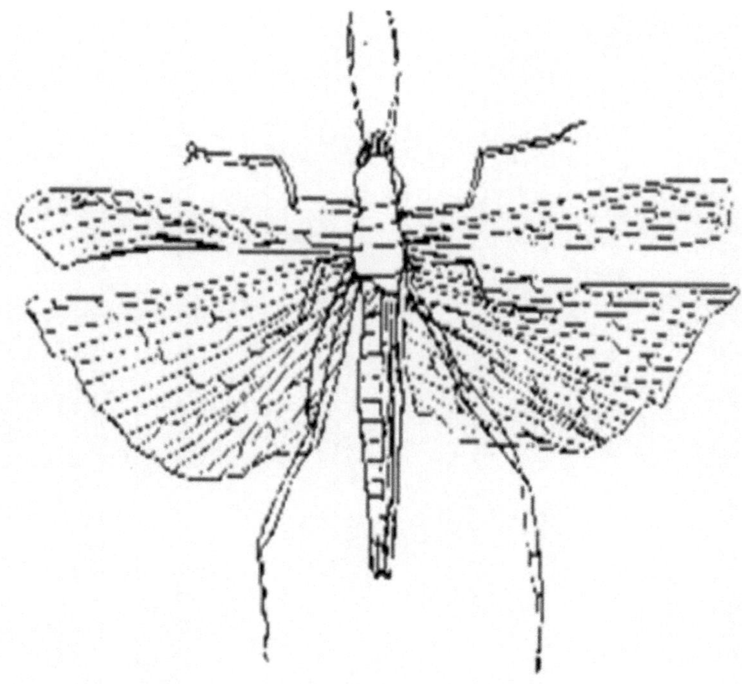

(A) Wings spread

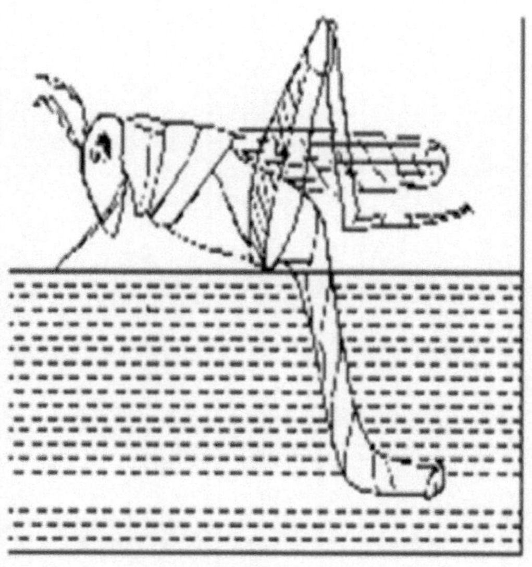

(B) Locusts in the act of oviposition

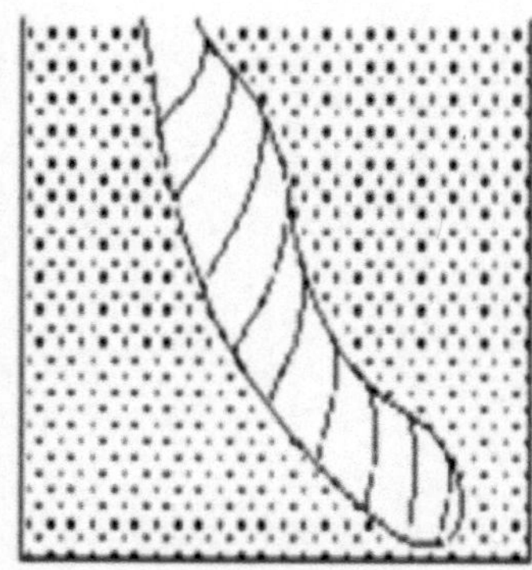

(C) Eggs in eggpod

17

Insect Pheromones and Hormones of Any Insect

DEFINITION

Pheromones are chemical produced as messengers that affect the behavior of other individuals of insects. The term *pheromone* was introduced by Peter Karlson and Martin Liischer (1959) and is based on the Greek *pherein* (to transport) and *hormon* (to stimulate) is a chemical that triggers a natural behavioural response in another member of the same species. There are alarm pheromones, food trail pheromones, sex pheromones. And many are others that affect behaviour or physiology. Their use among insects has been particularly well documented, although many vertebrates and plants also communicate using pheromones. Pheromones are chemicals released by an organism into its environment enabling it to communicate with other members of its own species.

TYPES OF PHEROMONES

Insect Pheromones and Alarm Pheromones

These pheromones are defined as chemical substances produced and released by an organism, that warm or afert another of the some species of impending danger. Some species release a volatile substance when attacked by a predator that can trigger flight (in aphids) or aggression (in bees) in members of the same species. When an ant is disturbed, it releases a pheromone that can be detected by other ants several centimeters away. They are attracted by low concentrations of the pheromone and begin to move toward the region of increasing concentration. As they get nearer to their disturbed nestmate, their response changes to one of alarm. The higher concentration causes them to run about as they work to remedy the disturbance. Unless additional amounts of the alarm pheromone are released, it soon dissipates. This ensures that once the emergency is over, the ants return quietly to their former occupations. Honeybees also have an alarm pheromone (which is a good thing not to elicit around a colony of "Africanized" bees). Examples, Melipona bees secrete citral pheromone and formicine ants secrete undecane.

Alarm Pheromones for ants and Queen Mandibular Pheromone of honeybees

Queen Mandibular Pheromone: Honeybee queens spend their lives literally surrounded by a retinue of worker bees. The workers are attracted to her by a pheromone 9-hydroxy-decinoic acid that she releases from here mandibular glands. The pheromone is a mixture of alcohols and organic acids. Among its effects:

1. Inducing the workers to feed and room her.
2. Inhibiting the workers from building queen cells and rearing new queens.
3. Inhibiting ovary development in the workers.

Sex Attractants: The pheromones are known with which one sex (usually the female) of an insect species attracts its mates. Many of these sex attractants - or their close chemical relatives - are available commercially. Examples, male cockroach secrete seducin, *Bombyx mori* secretes bombykol and *Musca nebulo* secretes musculure.

Male Confusion: Distributing a sex attractant throughout an area masks the insects own attractant and thus may prevent the sexes getting together. This *communication disruption* has been used successfully against a wide variety of important pests. For example, the sex attractant of the cotton boll weevil has reduced the need for conventional chemical insecticides by more than half in some cotton-growing areas.

Insect Monitoring: Insect sex attractants are also valuable in monitoring pest populations. By baiting traps with the appropriate pheromone, a build-up of the pest population can be spotted early. Even if a conventional insecticide is the weapon chosen, its early use reduces—

1. The amount needed
2. Damage to the crop
3. Cost to the grower
4. Possible damage to the environment.

Early detection of pest build-up is a key ingredient in the system known as integrated pest management (IPM).

Pheromone: Natural sex attractants

Releaser Pheromones of Mating Insects

These are powerful attractant molecules that some organisms may use to attract mates from a distance of 2 miles or more. This type of pheromone generally Ilicites rapid response but is quickly degraded.

Territorial Pheromones: These pheromones are laid down in the environment and mark the boundaries of an organism's territory. In dogs, these hormones are present in the urine, which they deposit on landmarks serving to mark the perimeter of the claimed territory. Example, male bumble bee secretes 2, 3-dihydro-6-trans-farnesol by its mandibular gland.

1. Fanning honeybee exposes Nasanov's gland (white – at tip of abdomen) Releasing pheromone to entice swarm into an empty hive 2. Aggregation of bug nymphs

Trail Pheromones of Ants: These pheromones are common in social insects. For example, ants mark their paths with these pheromones, which are non-volatile hydrocarbons. Ants synthesize pheromones in their pavan's and dufour's gland. Termites trail pheromone is secreted from sternal gland of 5th abdominal gland. Example, Zootermopsis secretes caproic acid. Some of the ants lay down an initial trail of pheromones as they return to the nest with food.

Sex Pheromones of Different Species of Inserts

In animals, sex pheromones indicate the availability of the female for breeding. Male animals may also emit pheromones that convey information about their species and genotype. Many insect species release sex pheromones to attract a mate and many lepidopterans can detect a potential mate from as far away as 20 km. Pheromones can be used in gametes to trail the opposite sex's gametes for fertilization. Pheromones are also used in the detection of oestrus in sows. Boar pheromones are sprayed into the sty, and those sows which exhibit sexual arousal are known to be currently available for breeding.

Ecdysone Hormone of Insects (PTTH) Prothoracicotropic Hormones

There are two prothoracic glands/ecdysial gland are ectodermal in origin which are located in the thorax and absent in apterygotes. Under the influence of PTTH, they secrete the steroid hormoney ecdysone. Acting together, PTTH and ecdysone trigger every molt: larva-to-larva as well as pupa-to-adult. It accounts for the dramatic changes of metamorphosis. Insects have rigid exoskeleton and can grow only by periodically shedding their exoskeleton – called molting. Molting occurs repeatedly during larval development. This marked transformation is called metamorphosis. Metamorphosis takes place during a dormant stage called the pupa. The sequence ending in the center panel (B) shows the larval, pupal, and adult stages during normal development of the domestic silkworm moth, Bombyx mori. Knowledge of insect hormones has provided a number of opportunities to enlist them or molecules related to them – in the battle against insect pests.

Juvenile Hormone (JH)

Juvenile hormone is secreted by two tiny glands behind the brain, the corpora allata. As long as there is enough JH, ecdysone promotes larva-to-larva molts. With lower amounts of JH, ecdysone promotes pupation. Complete absence of JH results in formation of the adult. Therefore, if the corpora allata are removed from an immature silkworm, it immediately spins a cocoon and becomes a small pupa, A miniature adult eventually emerges. Conversely, if the corpora allata of a young silkworm are place in the body of a fully-mature larva, metamorphosis does not occur. The next molt produces an extra-large caterpillar developed.

Insect Behaviour

Ethology is the scientific and objective study of animal behaviour. In 1930s and 1940s two European scientists, Konrad Lorenz and Niko Tinbergen developed the ethological strat of combining observation of animals in their natural habital with evolutionary analysis of the observed behaviour pattern. Ethologists are concerned with the biological significance of behaviour pattern and how these patterns have evolved. Many of the biologically important behaviour patterns studied by ethologists appeared to be primarily the product of genetic expression and less result of learning. The male rises in metathoracic leg on the side away from the female male turns away from the female. She may respond by placing her proboscis on the tip of male's abdomen. She then appears to pull the male backword along his perch in a spiral course.

Ethologists recognize the behaviour that an animal shows at any moment is the result of past selective forces. These determine a particular morphological biochemical and natural structure in an animal. The functional characteristics of this structure are continually altered by experience, hormones and other physiologic changes. The sum of these factors results in the maintenance of animal in its proper condition, and stimuli from the biotic and the abiotic environment can then elicit particular behaviour.

Toxic nectar affects behaviour of insect pollinators

Sexual conflict behaviour in insects can be seen in mating struggles between a male and female

What is the Behaviour of Insect

The behaviour is the action, which changes the relationship between the animal and its environment, as no animal lives alone in the biosphere or specific responses of a certain organism to a specific stimulus or group of stimuli. Each comes in contact with other animals during its lifetime. Even those animals whose reproductive needs can be satisfied in solitude do not live alone. They have contact with other animals for food. Therefore, animals depend on other animals for their survival. The survival also depends on the maintenance of suitable relations with nonliving environment. The behaviour may occur as a result of an external stimulus. Receptors are necessary to detect the stimulus. The nerves are needed to coordinate the response and the effectors carry out the potion. Behaviour may also occur as a result of an internal stimulus. A hungry animal searches for food. A thirsty animal behaves in a way, which will lead to satisfying its thirst. Therefore, the behaviour of an animal results from the combination of both external and internal stimuli. The internal stimuli, such as thirst and hunger, give motivation for the action.

TYPES OF BEHAVIOUR OF INSECTS

Behaviour types are determined by the capabilities of the organism's, effector and nervous system. When there is change in the environmental variable. The stimulus depends on how the receptor, nervous systems are interconnected and integrated. Their organization may be direct, so that the stimulus triggers a simple set of muscle action that proceeds automatically once started. As muscular action proceeds, it may be constantly monitored by other stimuli that it has generated in the body, which therefore, steer and guide it. The stimulus may trigger a complex response that is completely programmed genetically in the central nervous system.

Predatory Behaviour of Insects: In which one animal kills another for food, there is interaction between the attack patterns of the predator and the defense pattern of the prey. The predatorprey interaction shows some phenomenon like mimicry, crypsis and aposematic coloration.

Parasites Affecting Insect Behaviour

Mimicry: Mimicry is the close resemblance between one insect and another insect or a plant part.

1. **Batesian mimicry (Bates, 1862):** It is the harmless but is protected by virtue of its similarity to an unpalatable model, example Viceroy Butterfly (*Limenitis archippus*) which is a batesian mimic of monarch butterfly (*Danaus plexippus*).
2. **Wasmannian mimicry/Parasitic mimicry (Rettenmeyer, 1970)**: In these insects that live in close association with ants and have come to closely resemble their host, example, different types of ants that live together and are similar, and parasite resemble its host.

Parental Investment Theory of Insects

Paternal care is related to certainty of paternity. Natural selection acts against paternal care when paternity is uncertain. Paternal care should not include nuptial gifts and spermatorphores give to females to nourish the eggs. The complexities of paternal care are shown in three species of Belostomatidae (The Giant Water Bug). Maternal care is the most prevalent parental investment. Females provide nourishment and tissue from their bodies to provision the eggs. Guarding of eggs and nymphs from predators is found in many different orders. The Ovoviviparous leaf beetle (*Gonioctena sibirica*) stays with larvae until the last larval instar.

Social Behaviour of Insects: The Social insects, such as termites, ants and many bees and wasps, are the most familiar species of eusocial animal. They live together in large well-organized colonies that may be so tightly integrated and genetically similar that the colonies of some species are sometimes considered super organisms. It is sometimes argued that the various species of honey bee are the only invertebrates (and indeed one of the few non-human groups) to have evolved a system of abstract symbolic communication where behaviour is used to *represent* and convey specific information about something in the environment. In this communication system, called *dance language,* the angle at which a bee dances represents a direction relative to the sun, and the length of the dance represents the distance to be flown. Only insects which live in nests or colonies demonstrate any true capacity for fine-scale spatial orientation or homing. This can allow an insect to return unerringly to a single hole a few millimeters in diameter among thousands of apparently identical holes clustered together, after a trip of up to several kilometers' distance.

Learning Behaviour of Insects

Most insects (cockroaches) begin forgetting soon after they learn something whereas honeybees hold memories up to 14 days. In few insects selected to learn over long periods (life too short). Location of learning is found principally in the protocerebrum while some learning in the *mushroom bodies*. In cockroach it is found in thoracic ganglia in cockroach and called **ganglionic learning.** Insects can learn maze-learning, some can learn: ants and cockroaches and the built abilities are modest. Insect's fare better learning those things more closely tied to their survival like.

1. Honeybees learn (four) colors and various shapes,
2. ***Heliconiine* butterflies** (long wing butterflies) trapliners in tropics. Insect can also learn orientation flights; capture rapidly the complicated surface details and easily looled (*Ammophila*): move one or two key markers and all is lost while others not easily fooled [*(Bembix)* (digger wasp that nest on open sands)].

Innovative Approaches of Insect Control

Insect Repellent Approaches

An insect repellent is a substance applied to skin, clothing, or other surface which discourages insects (and arthropods in general) from landing or climbing on that surface. There are also insect repellent products available based on sound productions, particularly ultrasound. These electronic devices have been shown to have no effect as a mosquito repellent by studies done by the EPA (Environmental Protective Agency) and many universities.

1. Nepetalactone, oil also known as "catnip oil
2. Citronella oil
3. Permethrin ($C_{21}H_2O$ $C_{12}O_3$)
4. Neem oil
5. Bog Myrtle

For consumers trying to take a more environmental-health conscious approach to personal care products, the topic of bug spray can be a bit confusing.

Best mosquito repellent for your yard

Insect Repellents from Natural Sources

There are many preparations from naturally occurring sources that are repellent to certain insects. Some of these act as inseticides while others are only repellent.

- *Achillea alpina* (mosquitos)
- Alpha-terpinene (mosquitos)
- *Callicarpa americana* (Beauty berry)
- Camphor (moths)
- Carvacrol (mosquitos)
- Castor oil *(Ricinus communis)* (mosquitos)
- Catnip oil *(Nepeta* species) (nepetalactone against mosquitos)
- Cedar oil (mosquitos, moths)
- Citronella oil (repels mosquitos)
- Fennel oil *(Foeniculum vulgare)* (mosquitos)
- Lemongrass oil (*Cymbopogon* species) (mosquitos)

- Marigolds (*Tagetes* species)
- Neem oil (*Azadirachta indica*) (Repels or kills mosquitos, their larvae plethora of other insects including those in agriculture) •Oleic acid, repels bees and ants by simulating the "Smell of death" produced by their decomposing corpses. It is a 400 millions years old natural mechanism helping to sanitise the hives or to escape predators
- Pyrethrum (from *Chrysanthemum* species, particularly *C. cinerariifolium* a *C. coccineum*)
- Rosemary (Rosmarinus officinalis) (mosquitos)
- Solanum villosum berry juice (against *Stegomyia aegypti* mosquitos)

Advantages of Insects

1. It is safe to use.
2. It is non-toxic to man.
3. It can be used as spray by making synergist with pyrethrins.
4. Insects do not eat and die due to starvation.
5. It can be safely used on skin and most fabrics.
6. It is economic and not time consuming.

Disadvantages

1. Its efficiency depends on ability to deter greatest number of species from feeding for longest period.
2. All kinds of insects can not be repelled by the same chemical.
3. It does not kill the insect.
4. Insects able to return to their normal activities if repellent is losi as the repellents are active only for a short period.
5. They are required in large amount.
6. They are oily in nature and have disagreeable odor.

Insect Attractants and Traps

The insect attractants and traps are useful tools for monitoring insect populations to determine the need for control or the timing of control practices in some instances, attractants and traps also can be used to control insect population directly by mass trapping or mating disruption.

Using attractants and traps to monitor and control insect populations can improve the effectiveness of insecticide applications and sometimes reduce the use of broad-spectrum, more toxic compounds.

Light Traps

Light trap is a device used at night in the field to collect moths and other flying insects such as, Armyworm Bugs, Cutworm Flies. A great number of insect species are attracted to light of various wavelengths! Several species respond uniquely to specific portions of the visible and no-visible spectrum. They are attracted by fluorescent bulbs or bulbs that emi ultraviolet wavelengths (black lights). The species of moths, beetles, flies, and other insectsare attracted to artificial light. They may fly to lights throughout they night or only during certain hours. Key pests that are attracted to light include the European corn borer, codling moth, cabbage looper, many cutworms and armyworms, diamondback moth, sod webworm moths, peach twig borer, several leaf roller moths, potato leafhopper, bark beetles, carpet beetles, adults of annual-which grubs *(Cyclocephala)*, housefly, stable fly, and several mosquitos.)

Pitfall Traps

Insects that crawl about on the ground can be captured in a pitfall trap. The simplest trap can be constructed easily by placing a can or plastic container in the ground. Add enough killing agent (such as alcohol) to cover the bottom of the container. To keep rainwater out of these traps, a board can be propped up over the opening.

Advantages

1. All these attractants are mixed with food to make poison baits, trap baits etc.
2. They have no harmful effects, non polluted, and biodegradable.
3. It is specific in nature and do not effect non target species.
4. Insect do not develop resistance against attractants
5. Environment spreaded by attractants do not able the male to locate its mate (Male Confusing Technique).

Disadvantages

1. It is costly and time consuming.

2. Individual responds to stimuli according to relationship between the intensity of stimulus and their response threshold.
3. It varies according to physiological condition and may change with time.
4. Many insects have migratory phase during which attractants are less responsive to this control.
5. It produces variable results.

Biting Mites

Rat mites live in the nests of wild rodents. Mites are tiny arthropods, related to ticks. Several types ot mites can be found in homes and of these a few may bite humans. Most mites are harmless predators of insects, or feeders on decaying plant material. Some pest mites feed on stored products like cheese and grain. Others are merely nuisance pests, accidentally entering homes from their normal outdoor habitat. Only a few mite species are parasitic on birds or mammals, but these can occasionally become biting pests in homes. Identifying the type of mite and/or likely host is the first step in solving an indoor mite infestation.

Human Biting Mites

The several types of mites are associated with cases of skin dermatitis in humans. The tropical rat mite, *Ornithonyssus bacoti,* is one of the most common house invading species. The tropical fowl mite, *Ornithonyssus bursa,* and northern fowl mite, *Ornithonyssus sylwiarum,* are also frequently encountered in homes. The latter two species are found principally on domestic or wild birds. The house mouse mite, *Liponyssoides sanguineus,* may also be found in structures with house mouse infestations. The tropical rat mite is a parasite on rats. Although none of these species are truly parasitic on humans, they bite people readily, often producing dermatitis and itching

Rat and bird mite infestations occur in structures where rat or bird nests are located. Infestations are sometimes first noticed following determination, or after the natural hosts have died or left the structure infestations may also occur where heavy mite infestations have developed around a rodent or bird nest. Occasionally rodent or bird mites may be found on rodents kept as pets. Rat mites are small, approximately the size of the period at the end of this sentence. They move actively and can be picked up with a wet finger, brush or piece of sticky tape. Distinguishing between different species of *Ornithonyssus* mites to determine whether birds or rodents are the likely source is difficult and requires special expertise. The first course of action when faced with a suspected biting mite problem is to look for all potential bird or rodent sources and collect some of the mites, if possible.

1. The non-biting clover mite is reddish in color and has two long front legs 2. Rat mites live in the nests of wild rodents

Collecting mites: Most pest control companies will (rightly) not treat a home without proof of pest presence. It is therefore important to collect mites prior to treatment. Parasitic, mites are often first noticed when biting. Mites can be collected from the skin with an artist's brush or tissue dipped in rubbing alcohol. Mites collected in this way should be placed in a small vial or other waterproof container with a small amount of rubbing alcohol. Mites can also be collected from the skin with a piece of tape (although this makes accurate identification of the mite unlikely). Sticky traps are also useful tools for sampling tiny arthropods around the home. Place several sticky traps in rooms where bites are occurring.

1. Human spider mite 2. Mites are microscopic insects that travel through air and bite human skin.

Control: The primary mite *host* must be eliminated before successful control rodent or bird mites can be achieved. Clues to the type of host that has invaded the house can be deduced by the time of year that the mite infestation occurs. Rodent infestations are possible at any time of year, though they seen to occur most frequently in the fall and winter. Bird problems are most common during the spring and summer. Roof rats are the most common rat encountered in Texas homes as their name implies, roof rats are good climbers and often enter the home through openings in the roof or soffit areas. Noises in ceilings, especially at night, can indicate roof rat activity. Bird infestations are often first indicated by the sound of chirping coming from a chimney or soft area. The same rules and materials used for rodent-proofing are effective in keeping birds out of the home. Special screening may be needed on chimneys to deny bird's access to chimney areas. Birds nesting in chimneys may also indicate the need for chimney maintenance and cleaning. Chickens and other fowl kept in sheds or coops attached to a home can also be a source of mites indoors.

This should reduce the risk of live mites dispersing from the site and entering the indoor areas of the structure after the nest is removed. Long sleeves, gloves, and a tight-fitting dust mask are recommended when removing old bird or rodent nests to reduce the risk of exposure.

To ecto-parasites, like mites, and other pathogens

Other Mites: The non-biting clover mite is reddish in color and has two long front legs other mites that can be found in homes include the clover mite and certain mites associated with stored products. Clover mite infestations are common in homes during the late winter and early spring. Clover mites are feeders on grasses and weeds and can sometimes be found invading structures from the outside through windows and doorways. Adult clover mites are approximately 1 mm-long and can be distinguished, under magnification, by their long front pair of legs. These mites sometimes produce a red stain when crushed. Clover mites do not bite people, and are mainly a nuisance pest. Keeping grass and weeds cut short immediately around structural foundations, and maintaining tight seals around windows and doors can help reduce invasion of the home by this pest. Pesticide sprays can be applied to potential entry areas from the outside Indoor sprays are generally not needed for this pest.

The grain mite, *Acarus siro,* is one of the most common stored product mite pests this mite is found most frequently on processed cereal products (*e.g.*, flour); whole wheat flour is a preferred food source, as are some fungi and molds. Grain mites have also been recorded from cheese, poultry litter, and even from abandoned bee nests. Parasitic mites can be distinguished from stored product mites only with the use of a high-powered microscope; however the location of an infestation within a home may provide the best clue as to whether the pest is a feeder on stored grains. Most stored grain insects do not bite. Removal of the infested product, and thorough cleaning of the storage area, is usually sufficient to eliminate infestations. High humidity and moisture also favours mite infestations, so moisture control should also be a goal of a storage mite infestation. Rats are some of the most troublesome and damaging rodents in the United States. They eat and contaminate food, damage structures and property, and transmit parasites and diseases to other animals and humans. Rats live and thrive in a wide variety of climates and conditions and are often found in and around homes and other buildings, on farms, and in gardens and open fields.

Identification

People don't often see rats, but their presence is easy to detect. In California, the most troublesome rats are two introduced species, the roof rat and the Norway rat. It's important to know which species of rat is present in order to choose effective control strategies. Norway rats, *Rattus norvegicus,* sometimes called brown or sewer rats, are stocky burrowing rodents that are larger than roof rats. Their burrows are found along building foundations, beneath rubbish or woodpiles, and in moist areas in around gardens and fields. Nests can be lined with shredded paper, cloth, or other fibrous material. When Norway rats invade buildings, they usually remain in the basement or ground floor. Norway rats live throughout the 48 contiguous United States. While generally found at lower elevations, this species can occur wherever people live.

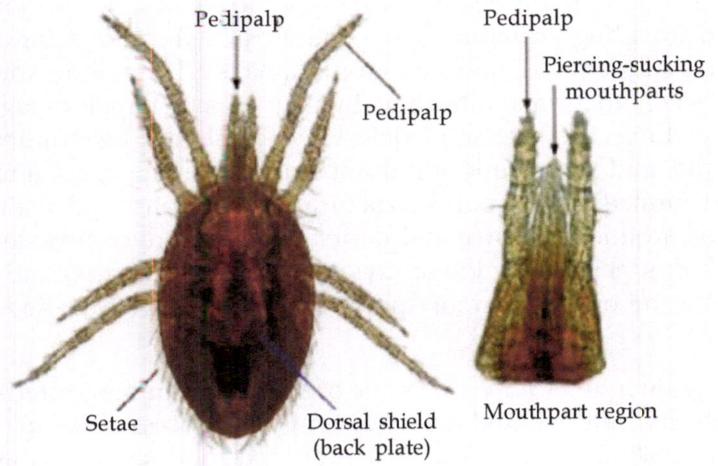

Mites that attack humans with some frequency

Roof rats, *R. rattus,* sometimes called black rats, are slightly smaller than Norway rats. Unlike Norway rats, their tails are longer than their heads and bodies combined. Roof rats are agile climbers and usually live and nest above ground in shrubs, trees, and dense vegetation such as ivy. In buildings, they are most often found in enclosed or elevated spaces such as attics, walls, false ceilings, and cabinets. The roof rat has a more limited geographical range than the Norway rat, preferring ocean-influenced, warmer climates.

Live Traps

Live traps aren't preferred, because trapped rats must be either humanely killed or released elsewhere. Releasing rat's outdoors isn't recommended, as they can cause health concerns to people, pets, and other domestic animals. Because neither the roof rat nor the Norway rat is native to the United States, their presence in the wild is very detrimental to native ecosystems.

They have been known to decimate some bird populations.

Rodenticides (Toxic Baits)

While trapping is generally recommended for controlling rats indoors, when the number of rats around a building is high, you might need to use toxic baits to achieve adequate control, especially if there is a continuous reinfestation from surrounding areas. If this is the case, consider hiring a licensed pest control applicator that is trained to use rodenticides safely.

Other Control Methods: Rats is wary animals, easily frightened by unfamiliar or strange noises. However, they quickly become accustomed to repeated sounds, making the use of frightening devices- including high frequency and ultrasonic, sounds—ineffective for controlling rats in homes and gardens. Predators, especially cats and owls, eat rats and mice. Some house cats don't have the ability or inclination to prey on adult Norway rats. Often, predators aren't able to keep rodent numbers below levels that are acceptable to most people further; pet food can serve as an attractant and provide a continuous food supply to rats and mice in suburban environments.

MITES: Mites belonging to the family Tetranychidae, commonly known as spider mites as they spin webs like spiders, are of immense economic importance as all are exclusively phytophagous and many are pests of a large number agricultural crops, fruit trees, vegetables, etc. often doing colossal economic loss to the growers. The common damage symptoms caused due to these mites are stunting of plant growth, severe defoliation, reduction in yield and often arious types of malformations and deformations of plant parts, fruit trees, etc Mites, along with ticks, aie small arthropods belonging to the subclass Acari and the class Arachnida. The scientific discipline devoted to the study of ticks and mites is called acarology. In soil ecosystems, mites are favored by high organic matter content and by moist conditions wherein they actively engage in the fragmentation and mixing of organic matter. Spider Illite **(Tetranychidae)** Spider mites are members of the **Acari** (mite) family Tetranychidae, which includes about 1,200 species. They generally live on the undersides of leaves of plants, where they may spin protective silk webs, and they can cause damage by puncturing the plant cells to feed. Spider mites are known to feed on several hundred species

Sub-class		Acari
Class	:	Arachnida
Order	:	Trombidiforme's
Superfamily	:	Tetranychoidae
Family	:	Tetranychidae

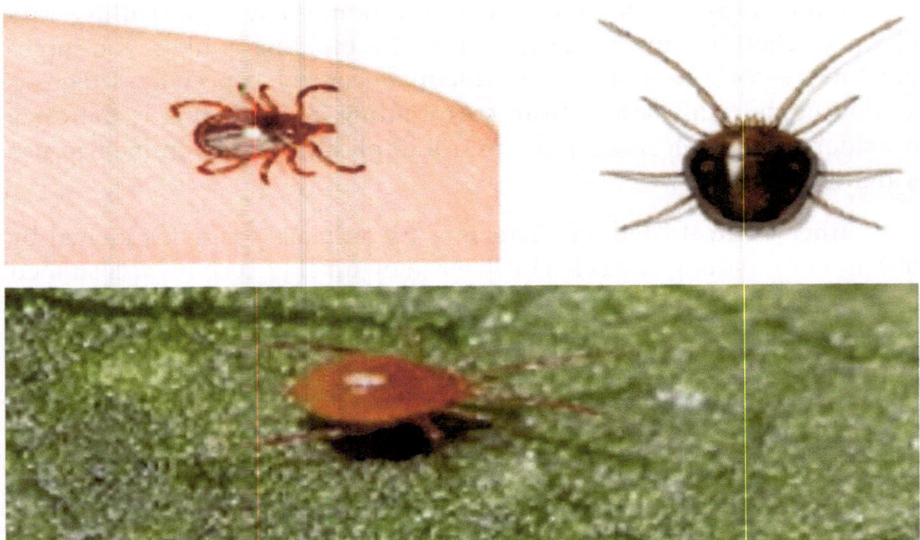

1. Poultry and Red Mite 2. Picture of what Bird Mites look like

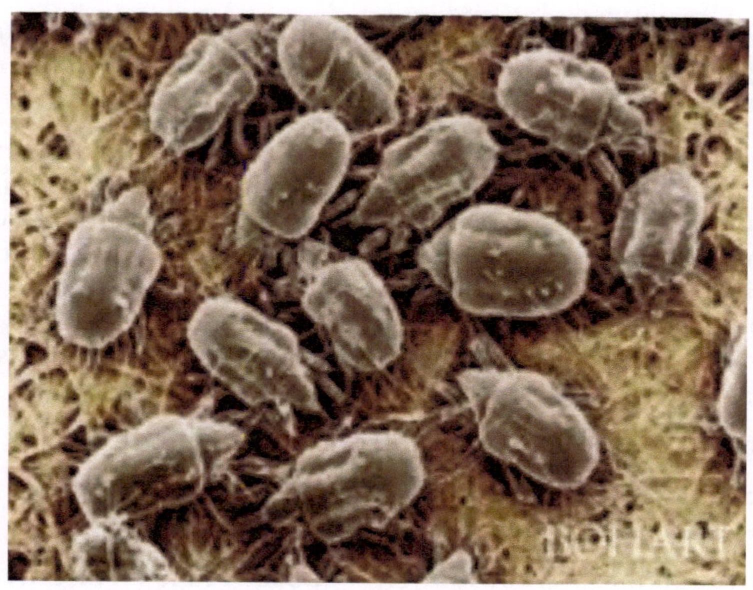

Human Skin Parasites

All about Spider Mites and Cannabis Plants

Spider mites are less than 1 millimetre (0.04 in) in size and vary in color. They lay small, spherical, initially transparent eggs and many species spin silk webbing to help protect the colony from predators; they get the "spider" part of their common name from this webbing

Life-cycle is hot and dry conditions are often associated with population build-up of spider mites. Under optimal conditions (approximately 80°F or 27°C), the two-spotted spider mite can hatch in as little as 3 days, and become sexually mature in as little as 5 days. One female can lay up to 20 eggs per day and can live for 2 to 4 weeks, laying hundreds of eggs. This accelerated reproductive rate allows spider mite populations to adapt quickly to resist pesticides, so chemical control methods can become somewhat ineffectual when the same pesticide is used over a prolonged period. Spider mites, like hymenopterans and some scale, insects, are arrhenotochous: females are diploid and males are haploid. When mated, females avoid the fecundation of some eggs to produce males. Fertilized eggs produce diploid females. Unmated, unfertilized females still lay eggs that originate exclusively haploid males.

Damage to Crop: Spider mites generally feed on the underside of leaves, but will cover the entire leaf surface when populations are high.

They pierce plant cells and withdraw the cell contents. Feeding results in small clumps of dead cells and a speckled appearance of infested leaves. Wilting, leaf deformity, desiccation, and abscission occur with prolonged, high density infestations. Disruption of photosynthesis results in plant stunting and reductions in yield.

Management: Spider mites are often considered a secondary pest with damaging populations frequently occurring after application of broad spectrum insecticides (particularly sevin and most pyrethroids). A growing set of data also indicates increased problems with spider mites following soil applications of neonicotinoid insecticides. While use of these products is frequently recommended for other pest situations. When spidermites are present in the field, potential influence on mites should be considered in the decision.

Common name: Two spotted spider mite.

Scientific name: *Tetranychus urticae* Koch (Arachnida: Acari: Tetranychidae).

Introduction – Distribution – Description – Life-Cycle – Economic Importance – Management – Selected References.

Introduction

The two spotted spider mite. *Tetranychus urticae* Koch, has been controversial in its taxonomic placement. About 60 synonyms included under this species have compounded the controversy. The body of a spider mite is separated into two distinct parts: (1) the gnathosoma and (2) the idiosoma. The gnathosoma includes only the mouthparts. The idiosoma is the remainder of the body and parallels the head, thorax and abdomen of insects, after hatching from the egg, the first immature stage (larva) has three pair of legs. The following nymphal stages and the adult have four pairs of legs.

Two spotted spider mites, *Tetranychus urticae* Koch

18

All Insect Pests of Agriculteral Crops

COMMON INSECT PESTS AND THEIR CONTROL

Since, insects are the most successful and one among the strongest rivals of man in the competition of existence. Due to their immense tendency of producing numerous viable propagules (eggs) *i.e.*, their high fecundity, their population graph takes a sharp peak in a very short span of time.

A time reaches when the population of an insect type reaches to the level where from it starts causing inconvenience or injury to man, plants, domesticated animals and other human belongings, at such a juncture they are termed as insect pests.

Many, workers are of the view that "insects capable of causing 5 or more than 5% of damage to the crops, stored products, domesticated animals and other human dependences will be termed as pests". Thus, if an insect causes a loss of less than 5% will not be considered as a pest. Moreover insect pests which cause a damage of 5-10% are said to be minor pests and those which cause a loss of more than 10% of yield are said to be major pests.

On the basis of their feeding habitat they may be classified as under:

1. **Monophagus insects:** As the name it indicates that these insects are strictly confined to a single species of plants and feed only on a group of closely related plants. Simply we may say that a monophagous insect damages a single species of plants and its closely relatives. Thus, such an insect could be pest for one particular species and neautral or normal insect for other plant species or even useful in many cases. A typical monophagous insect is silk worm of mulberry.
2. **Oligophagous insects:** These insects feed on a group of closely related plant species mostly belonging to the same genus or family

consider the example of potato tuber moth that feeds only on the members of solanceae. Mustard saw fly and diamond black moth are confined to the plants falling in family. *Brassicaceae or cruciferae.*

3. **Polyphagous insects:** These insects feed on diverse plant species. Thus, they can be considered advanced insect forms due to their less dependence on a specific plant species. Their versatile feeding habitat has made them economically important pests to be studied. Some of the notable polyphagous insects are locusts, grasshoppers, termites, gram pod borer etc.

Pest Outbreak

The already stated that insects are known as pests when their population increases to such an extent so that they cause economic losses. This increase in the population of insect species is referred as pest outbreak; it could be brought about by following conditions:

i) *Favourable weather conditions:* The favourable weather conditions may lead to rapid production of an insect either directly or indirectly. Primarily favourable weather enhances their reproductive potential and side by side these conditions prove hazardous to their enemies, hence increasing their chances in the struggle for existence.

ii) *Deforestation:* A large group of insects are known to occupy forests as their habitats, But due to deforestation man has forced them to migrate on cultivated crops where they are free from their natural enemies which adds to their reproductive potential. Hence deforestation has also proved unfavourable agent for crop production by helping pests to grow in their number and conversery decreasing their yield.

iii) *Destruction of natural enemies:* To maintain ecological balance nature nurtures its creature so that they feed on one another. The birds, snakes, frogs, lizards, etc. are well known natural enemies of insects, they feed on them to satisfy their needs and strengthen the natural balance of an ecosystem too. Due to various man made activities the fecundity of these natural enemies has decreased which in turn has led increase in the pest population. The outcome of such activities has already been tasted by man. In 1956 their was a mass killing of sparrows in China which resulted in enormous increase in pest population so that Chinease Agriculture Ministry was forced to use insecticides via helicopters for crop rescue.

iv) *Introduction of new pest:* Due to cross border trades in terms of food products there is always some possibility that respective pests may reach to the lands were their hosts (food items on which they feed) are taken possibly in new land they are free from their enemies, thus, they rapidly increase their number. One of the intresting examples of such a case is the introduction of potato tuber moth in India from Italy. It came with the Italian potatoes and has become the regular pest of India now.

v) *Frequent use of pesticides:* The immense use of pesticides has unfortunaly proved to be an enhancing factor of pest outbreak. Frequent use of pesticides increases the number of pests in following ways.

   a) Since their natural enemies die and in the coming seasons they breed without any natural selection pressure.

   b) In many cases these pesticides kill other insects which are the natural competitors for targeted pests, hence their use favours pest's outbreak by removing their competition and their chances for existence increase exponentially.

   c) Frequent use of insecticides cause selective eradication of insects susceptible to insecticides so as to leave only the resistant ones to breed unaffected. These resistive forms usually grow fast and remain almost unaffected upon normal doseing.

Pest outbreak could be brought by some other factors too like:

i) Easy availability of food.

ii) Presence of large number of females than males in a population.

iii) Shortened life-cycle, so that many generations of insects are produced in a short span of time.

iv) Sudden increase in reproduction potential of an insect.

v) *Monoculture.* it is defined as the condition which involves culturing of one kind of crops contineously over a given piece of land. It lets insects thrive well by producing limited competition and easy availability of food.

Pests of Rice

1. **Yellow Stem Borer,** Local Name: Tana bedhak

   Systematic Position:

Order	:	Lepidoptera
Family	:	Pyraustidae

Genus : Scirpophaga (Tryporyza)

Species : incertulas

Host: Rice seems to be the only plant attacked by this pest and none of its alternate host is known till date.

Distribution: This pest is common in almost all the Asian countries. In India it is reported from Assam, Bengal, Odisha, Uttar Pradesh, Punjab, Andhra Pradesh and Tamil Nadu.

Life history: The entire life-cycle of this pest passes through following four stages *i.e.*, egg, larva, pupa and adult.

Egg stage: The moths become active after dusk when they mate and lay eggs on the underside of the leaves in 2-5 clusters of 60-100 eggs each. These eggs are covered with buff coloured hair of the female tuft. They hatch in 6-7 days and small black headed caterpillars soon bore into the stem from the growing points downwards.

Larva: The larva grows in 6 stages and is full-fled in 15-30 days initially they enter the leaf sheath and feed upon the green tissue of the stem for 1st 2-4 days. Thereafter they make their way into the stem near the node by boring. Sometimes wind takes them away so that they infect other surrounding plants A fully grown larva is about 20 mm which is white or yellowish white in colour with brown head and pronotum. The winter-born larvae dipause in this condition untill favourable condition are retained.

Pupa: Pupation takes place mostly inside the stem but sometimes it also takes place in straw or stubble. It should be noted that the larva before going to pupate constructs an emergence hole which is always located above the water level and pupates inside the attacked plant. The larva prepares a silken lined chamber and a silken cocoon around itself. The pupa is white yellowish in colour and measures about 14 mm in length and 2.5-3 mm in width. The pupal stage last for a week period sees in figure.

Adult: The adults produced after pupation are well distinct. The female moths are larger than the males with bright yellowish wings having minute but remarkable black spots. The male is light brown in colour with a large number of small brown dots along the sub terminal area. Moreover the females have a prominent tuft of brownish yellow silken hair at the tip of their abdomen.The number of generations vary from place to place its 5 in Tamil Nadu, 4 in Punjab 3 in Odisha and Bengal.

Damage: This pest has proved heavy hazardous to the basmati varieties than coarse varieties. Since the larva feeds inside the stem resulting in drying of the central shoot. In older plants of rice it causes drying of the panicle or white ear.

Preventive measure and control

i) Close planting and continuous stagnation at early stages should be avoided.

ii) The removal of stubbles and their destruction at the time of ploughing.

iii) Repeated ploughing and flooding could also prove effective measures in killing the larvae.

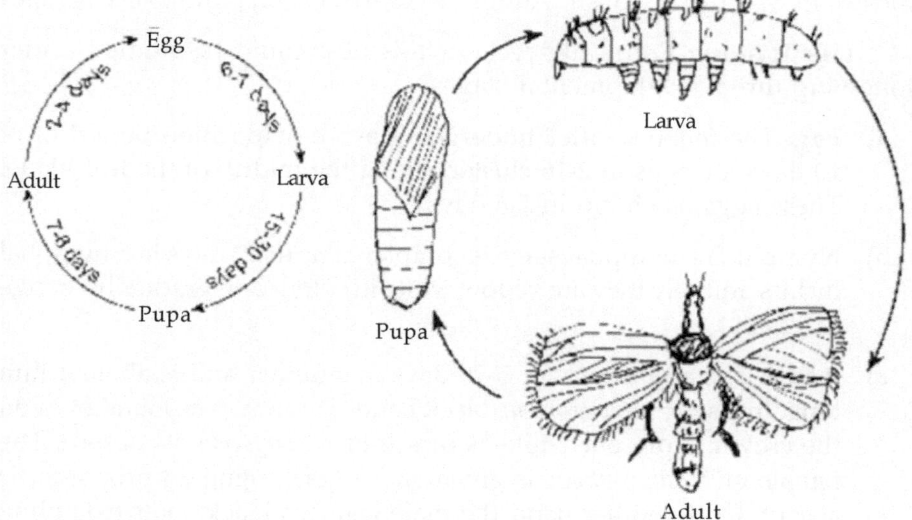

Life-cycle of *Scirpophaga incertulas*

iv) Collection of egg masses and their destruction.

v) As already mentioned that the eggs of this pest are laid near the tip of leaf so clipping of tips of seedings before the are transplanted in the rice fields can reduce their population.

vi) Use of light traps to attract the moths followed by their killings.

vii) Egg parasitoids like *Trichogramma Japonicum* and *T. Chilonis,* their simultaneous release starting after 30 days of transplantation reduce the borer damage.

viii) The crop may be sprayed by 875 ml of triazophos 40EC or 1.4 ltrs of monocrotophos @ 1000 ml/hectare.

ix) Use of resistant varieties like IR 20 and IR 26 may be used as alternatives.

2. **Green Leafhopper** Local Name: Dhan ka Phudka

Systematic Position

Order	:	Hemiptera
Family	:	Cicadellidae
Genus	:	Nephotettix
Species	:	*Virescens nigropictus impectieps*

Host: Rice, in its absence grasses act as alternative food.

Distribution: It is found in all the rice growing regions of India mostly in Madhya Pradesh, Andhra Pradesh, Odisha and West Bengal.

Life History: Entire life-cycle of this pest could be studied under following three developmental stages:

a) **Egg:** The females, after undergoing a pre-oviposition period of 6-10 days lay eggs in 2-16 clusters along the midrib of the leaf blade. These eggs hatch out in 2-6 days.

b) **Nymph:** The nymphal stage is completed in 10-22 days in 5 nymphal instars. Initially they are yellow white in colour and gradually change to green.

c) **Adult:** The adult lives for 5-20 days in summer and is about 4 mm long. Incase of *N. nigropicus* black band between compound eyes on the crown is present while as incase of *N. virescens* its absent. The female of *N. nigropticus* is green and a black tigne on pronotum is absent. On the other hand the male has two black spots extending upto the black distal portion on the forewings. There are about 6 overlapping generations from March to November and the pest is most active during July to September every years.

Damage: The adults and nymphs of this pest feed on the sap of rice they suck it from the leaves of plant that ultimately turn brown. The yield decreases to a large extent because plant looses its vigour.

Preventive measure and control

i) Close planting should be avoided.

ii) Alternate drying and wetting the field during peak infestation and draining out the standing water from the field 2-3 times helps in checking the population of this pest.

iii) Apply neem cake about 12 kg/20% nursery as a basal dose.

iv) Use of resistant varieties like IR 50, PB 18 remains almost unaffected from this pest.

v) Use of insecticides like Endosulfan 35 EC-1000 ml/ha per ha. Monocrotophos 36 SL-1.4 ltr/250 ltr of water. Chlorpyriphos 20 EC-2.5 ltr/250 ltr of water per ha. Fenthion 100 EC-500 ml/ha

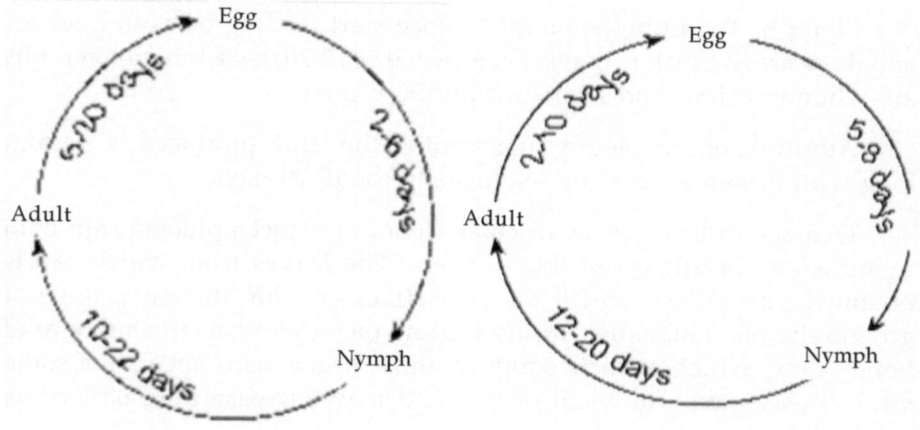

Life-cycle of *Nephotettix virescens*

3. **Brown Planthopper** Local name: Phudka, Bhusadi

Systematic Position

Order	:	Hemiptera
Family	:	Delphacidae
Genus	:	Nilaparvata
Species	:	*iugens*

Host plant: Rice but sometimes few grass species also act as host.

Distribution: It is the most destructive pest of rice in south and south-east, Asia, China, Japan, Indonesia, Korea, Sri Lanka, Taiwan, Thailand, and Phillipines, Malaysia etc. In India it has occupied the rice fields of Uttar Pradesh, Madhya Pradesh, West Bengal, Tamil Nadu, Andhra Pradesh, Haryana and Punjab. It is economically important to study as it has become the serious pest on high yielding varieties of rice.

Life History: Following three developmental stages are found in the life history of this pest.

Egg: The eggs are laid by the female with 3-10 days of their emergence in clusters of 3-14 by lacerating the parenchymal tissue along the mid rib of the leaf. Eggs are laid in masses and there are about 2-10 eggs/mass and a females lays about 125 egg masses. These eggs are some what dark and cylindrical, having two well distinct spots. It takes 5-8 days for eggs to be hatched out.

Nymph: The numphs on emergence start feeding on young leaves and these are five instars that are completed in 12-20 days. Initially nymphs are white in colour and later turn purple brown.

Adult: After completing five months, the adult produced is 4-5 mm long with brown eyes. They live usually for 10-20 days.

Damage: It is a typical vascular feeder and sucks phloem sap; both nymphs and adults cause this damage. The leaves from which sap is obtained turn yellow and if the pest attacks during the early stage of growth the plant may die. During feeding they excrete a large amount of honey dew, which attracts sooty moulds. It acts as a vector for some viral diseases too. The yield of the field may decrease upto 60% of its production.

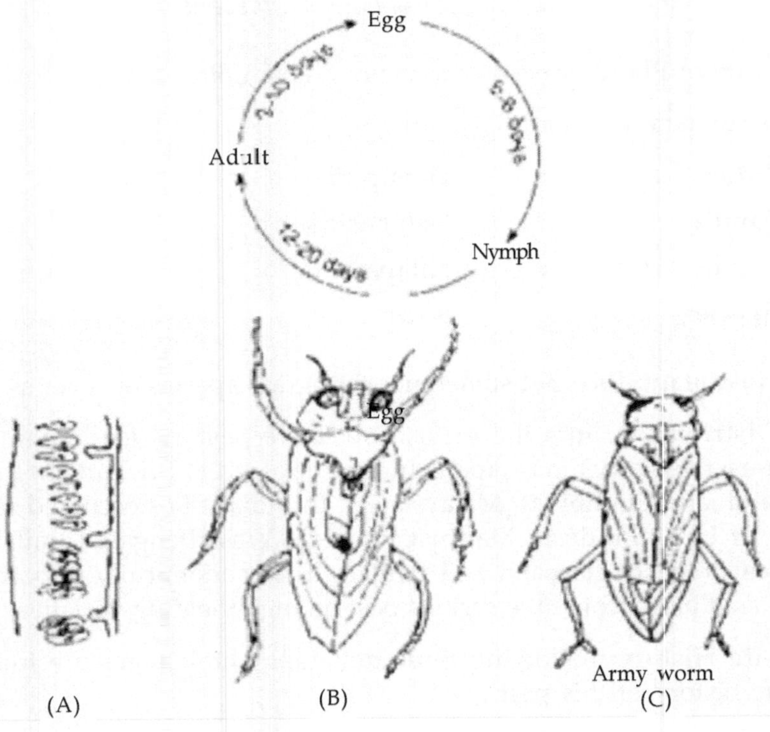

Life-cycle of *Nilaparavata iugens*

Preventive measures and control

i) Avoiding close planting.

ii) Avoiding excessive use of nitrogenous fertilizers.

iii) Alternate drying and wetting of the field during peak infestation should be done.

iv) Resistant varieties like Asha, Aruna, Co 42 etc. should be grown.

v) Regular doses of following pesticides should be directed towards the base of the plants:

a) Endosulfan 35 EC-100 ml/ha.

b) Methyl demeton 25 EC-1000 ml/ha. (*c*) Dichlorvo 76 WSC–350 ml/ha.

Repeat application if hopper population persists even after first application.

4. **Army Worm Common** Local Name: Katra, Sainik Keet Systematic Position:

Order	:	Lepidoptera
Family	:	Noctuidae Indian armyworm
Genus	:	Mythimna *(Persectania dyscrita)*
Species	:	*convecta*

Host plant: Paddy, maize, wheat and some grasses.

Distribution: The insect occurs in wheat and rice fields of Assam, West Bengal, Odisha, Bihar, Punjab, Andhra Pradesh and Peninsular India.

Life history: The insect completes its life-cycle in following four developmental stages.

Egg: The eggs are laid in overlapping rows on dry or green leaves or sometimes on soil. The egg period lasts for 3 days during which a single female lays 500-700 eggs. These eggs are small and spherical in shape which are initially light green in colour and turn pale yellow to black in later stage.

Larva: The newly hatched larva is quite active and turns from dull white to green colour. A full grown larva is about 36 mm long with nocturnal habit. There are 3 stripes on the lateral side of larva. The head is horny combed with dark lines and each proleg is provided with dark band on its other side and a dark tip on the inner side. There are six larval instars and larval period last for 20-30 days.

Pupa: The fully grown larva pupates in dumps of paddy, in cracks and cervices in ground or in loose soil. In the late instar, the larva spins a cocoon. Pupa is dark brown in colour and about 13 mm long. The pupal stage lasts for 9-14 days.

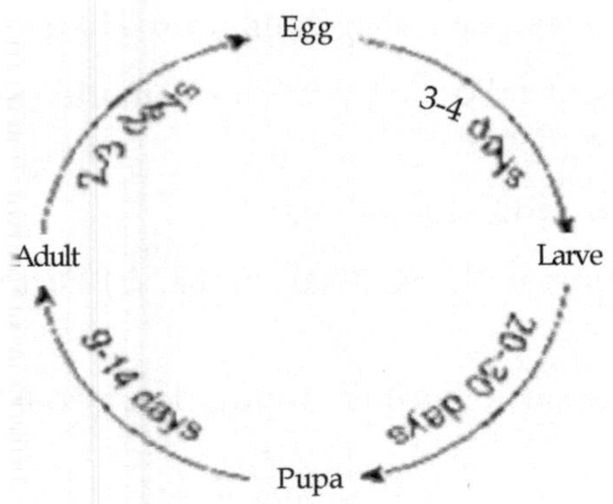

Life cycle of Mythimna

Adult: The adult is pale-brown moth measuring 40 mm and is gregarious in habit. They remain hidden during day time and become active at night.

Damage: The newly hatched larvae feed on the epidermis of the tender leaves. It is a destructive polyphagous pest and they damage to the paddy leaves and ears (also called Rice ear-cutting caterpillar) caused by the caterpillars. The damage to the paddy crop is caused mostly during September to November. They are voracious feeders and migrate from one field to another at rapid pace.

Preventive measure and control

i) The field should be flooded to kill the larvae.

ii) Proper spacing should be maintained while transplanting.

iii) Collecting the caterpillars followed by their destruction could prove an efficient means to check their population.

iv) Dusting the crop with chemicals like methyl parathion 2%-25 kg/ha.

v) Spraying the crop with endosulfan 35 EC – 750 ml/ha quinalphos 25 EC – 100 ml/ ha. Nozzle should be directed at the base of the clump.

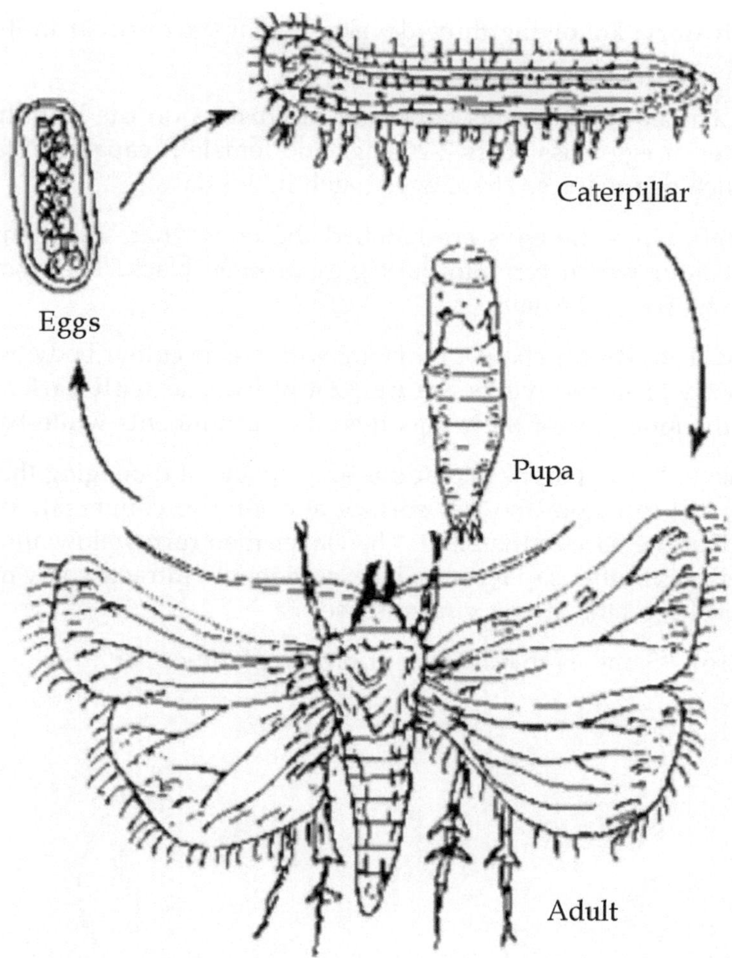

Life-cycle of *Mythimna* (separata) *convecta*

5. **White-Backed Plant-hopper** Local Name: Gudia, Phudake

Systematic Position:

Order	:	Hemiptera
Family	:	Delphacidae
Genus	:	Sogatella
Species	:	*furcifera*

Host plant: Rice seems to be its main host and grasses form the alternate one.

Distribution: It is a serious pest of rice in India, Pakistan, China, Sri Lanka and Myanmar. It is founds almost all over the country but its serious pests of southern states.

Life history: Following three-developmental stages occur in its life-cycle.

Egg: The adults lay eggs generally in clusters on the leaf sheath. Each cluster or egg mass bears 2-20 eggs and female is capable of laying 100-110 such egg masses. These eggs hatch in 2-5 days.

Nymph: Once the eggs are hatched the newly hatched nymph is white in colour which turns to dark grey or even black. The nymphal period is last for 10-12 days.

Adult: The adult is about 4 mm long with cream colour body, wedge shaped body. The forewings are uniformly hyaline with dark veins. Between the junctures of the wings there is a prominents white band.

Damage: Both nymphs and adults are capable of damaging the crop they suck cell sap from the leaf surface and tend to congregate on the leaf sheath at the base of the plant. The leaves then turn yellow and later on rust red. Like other sap feeders, its excretion also attracts sooty mould but its not the vector of any viral disease.

Control: "Same as that in case of brown planthopper".

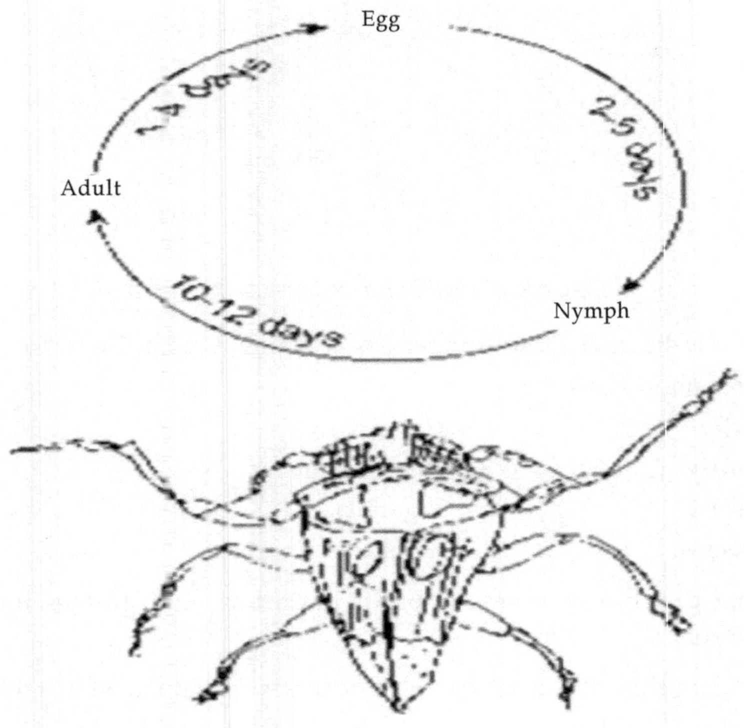

Sogatella furcifera

6. **Rice bug** Local Name: Rainya, Gundhi, and Chusia Systematic Position:

Order	:	Hemiptera
Sub order	:	Heteroptera
Family	:	Coreidae
Genus	:	Leptocorisa
Species	:	*variconis; acuta*

Distribution: It is found in many Asian countries like China, Japan, and Sri Lanka etc. It is also found in Australia and Orient. In India it is found in all major rice cultivating states like Uttar Pradesh, Bihar and Punjab.

Life history: Following three-developmental stages are found in the life history of this pest.

Egg: Eggs are laid on the lower side of leaves in rows. Eggs are round yellow in colour. They take about 5-7 days to hatch out.

Nymph: The nymphal stage lasts for two to three weeks during which they undergo six developmental phase to produce adults. The nymphs are immediate juice suckers and a fully grown nymph is about 13-16 mm long.

Adult: The adult is slender about 20 mm long and greenish brown in colour. It lives for 33-36 days and many generations are completed in a year. Adult has long legs and jointed antannae of red colour.

Host: Rice acts the main host however; it also feeds on maize, millets, sugarcane and on some grasses.

Damage: It is one among the most influential pests of rice known for its bad smell that it emits and got its name as 'gundhi' bug. Both nymphs and adults suck plant juices. During grain formation stage, this pest sucks the milky juice and leaves the chaffy husk. Black or brown spots appear around the holes made by the bug on which sometimes sooty mould may develope. The insect is very common in rice fields from August to mid November.

Control or Management

i) Use of light traps for collecting nymphs and adults.

ii) All other weeds and surrounding grasses should be completely destroyed.

iii) Dust carbaryl 5% or Malathion 5% @ 25 kg/ha.

iv) Use of resistant varieties could prove beneficient like Sona, Sathia etc.

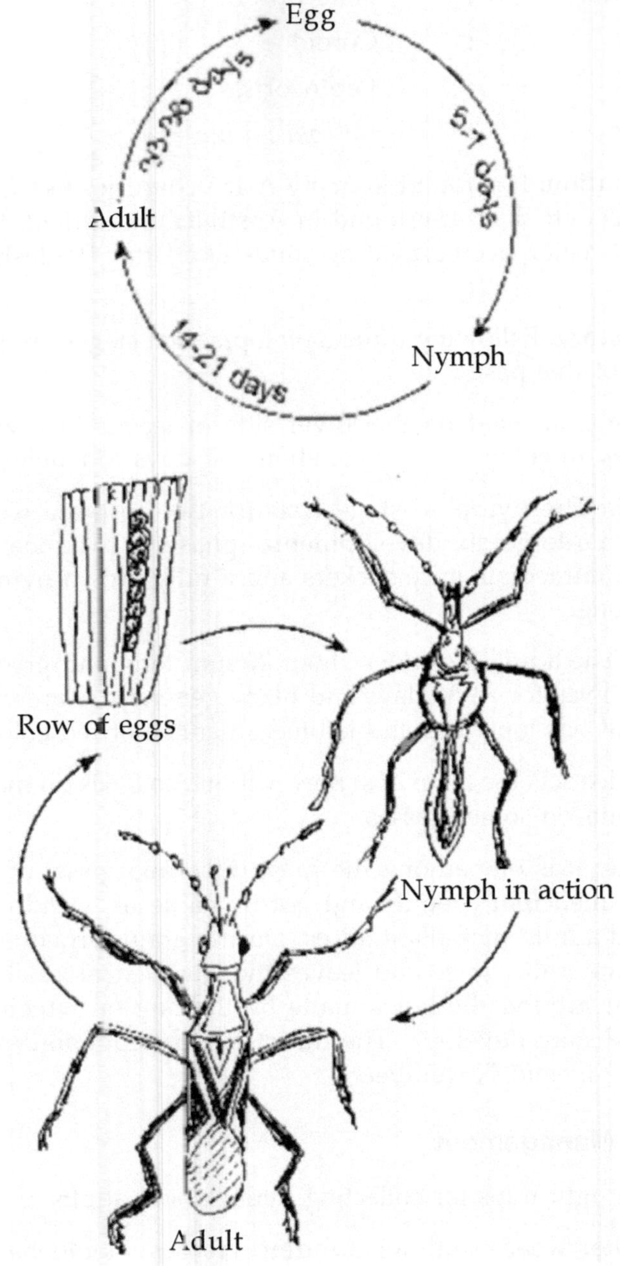

Life-cycle *Leptocorisa variconis*

Pests of Maize

India is one among the largest maize producing countries in the world with its annual production of 17300000 tonnes. Therefore their pests are economically important to study.

Some of the important pests of maize are summarised below:

1. Maize Stem Borer Local Name: Mokora, Makaya, Khera, Ratwa etc.

Systematic Position

Order	:	Lepidoptera
Family	:	Crambidae
Genus	:	Chilo (moth)
Species	:	*partellus*

Host: Maize, bajra, sugarcane, grasses etc. **Life cycle.** Entire life history of this pest could be studied under following four developmental stages.

Egg: After two days of copulation the female starts laying eggs in clusters of 500-100 in rows on the lower surface of leaves which are yellowish or creamy white in colour. Hatching period lasts for 3-6 days.

Larva: The newly hatched larva feeds on the leaves, making a few shot holes and then bore their way downwards through the central whorl as it opens. The freshly hatched ones are about 2 mm long with black head and prothorax. Larval period is of 21-28 period though it may extends even 150 days during winters. The larva becomes full fed in 14-30 days passing through six stages and after making a hole in the stem so that it could pupate there.

Pupa: Pupation occurs inside the special chambers constructed by the boring larvae. Pupa is reddish-brown in colour and pupal stage lasts for 5-8 days and then moths emerge out through a hole already constructed by the larva, before that has gone pupation. It is worthy to note that female pupa is some what larger and broader than male pupa.

Adult: The adult moth is a medium sized insects with a wing expanse of 3 cm. Forewings are light brown or straw coloured with numerous shinning brown spots on the margin and the hind wings are white and papery. Once the moths coupulate the male dies soon where as female dies after 2-3 days of egg laying.

Damage. To maize maximum damage is caused in the month of August. The damage is done by the caterpillars by feeding inside the stem and producing 'dead hearts'. Initially larvae begin to feed on the tender leaves there after they make their entry into the shoot through the mid whorl. The infected plants show stunted growth and does not produce grains.

Control or Management

i) Ploughing the field after harvest so that stubble weeds and other hosts are completely destroyed.
ii) Destruction of crop residues.
iii) Removal and destruction of dead hearts.
iv) Showing cowpea as an intercrop.
v) Use of insecticides like phophamidon

85 Sl-200, 250 and 300 ml/ha. at an interval of 2, 5 and 7 weeks of sowing Respectively.

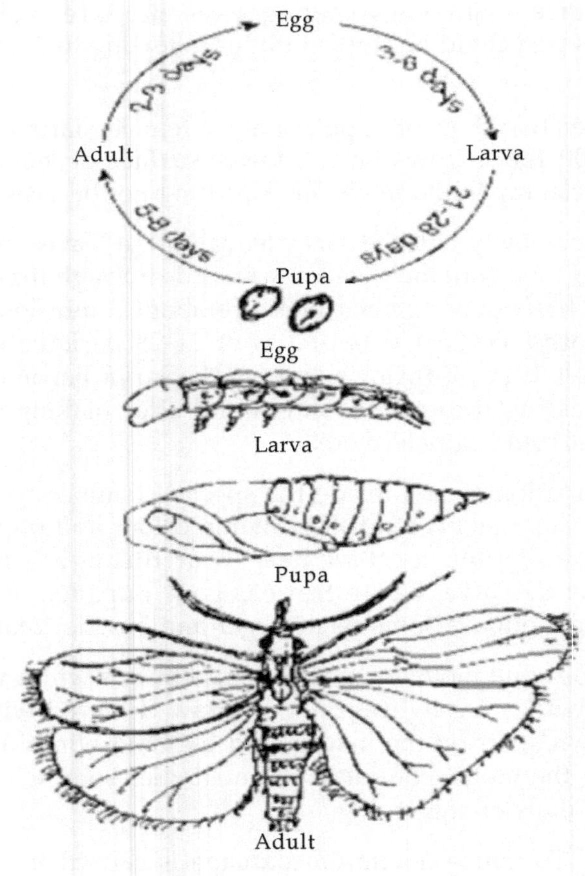

Life-cycle of Chilo partellus

2. **Asian Maize Borer**

Systematic position

Order	:	Lepidoptera
Family	:	Crambidae
Genius	:	Ostrinia
Species	:	*furnacalis*

Host plant: Its main host is maize however it also feeds on sorghum, millets and other grasses like *Artemesia*.

Distribution: Its distribution extends from China to Australia. Its is one of the worst corn pest in Japan and China. It's also found in Malaysia, Myanmar, Thailand, Korea, Vietnam, New Guinea and Micronesia. In India it occurs in many corn fields of Madhya Pradesh and West Bengal.

Life-cycle: Life history of this moth is completed in following developmental stages.

Eggs: Eggs are laid by the female moths which are light yellow or cream color and turn black as the time for hatching approaches. Eggs are laid on the underside on the upper leaf or husk and hatch out in 3-4 days.

Larva: The newly hatched larvae feed within the leaf whorl, under a protective silken web. There are generally 6-larval instars. Young larva is pink or a yellowish-gray in colour. Mature larva is white black spots on each segment. After about 9 days of feeding, they migrate downwards and bore into the stem. They feed there and complete their remaining instars in about 25 days.

Pupae stage: Normally, pupation takes place inside the tunnels made in the stem but they are capable of pupation inside the maize ear or in the stalk of ear. Pupation usually lasts for 6-9 days.

Adult: Adult female moth is yellow to light brown in colour with about 28 mm wing expanse. Males are slightly darker than females and have a tappering abdomen. Adults can live for 10-12 days and are prolific fliers.

Damage: This pest usually attacks the crop after one month of sowing. The larvae firstly feed on the soft tissues of leaf and then invade the midrib of the leave through which. It bores and ultimately reaches into the stalks and ears. In many conditions the pest attacks the base of the stalk so that plant collapses and even dies. It heavily affects the yield and suffered plants fail to recover.

Control or Management

i) The stubbles should be collected and destroyed, when harvesting is over.

ii) Use of light traps for collecting the insects. (*iii*) Sowing intercrops like cowpea could minimize the damage. (*iv*) Use of following insecticides have been found effective.Carbaryl 50 WP-1.0, 1.5 and 2.0 kg/hec. Endosulfan 35 EC-1.0, 1.25 and 1.5 lit/he

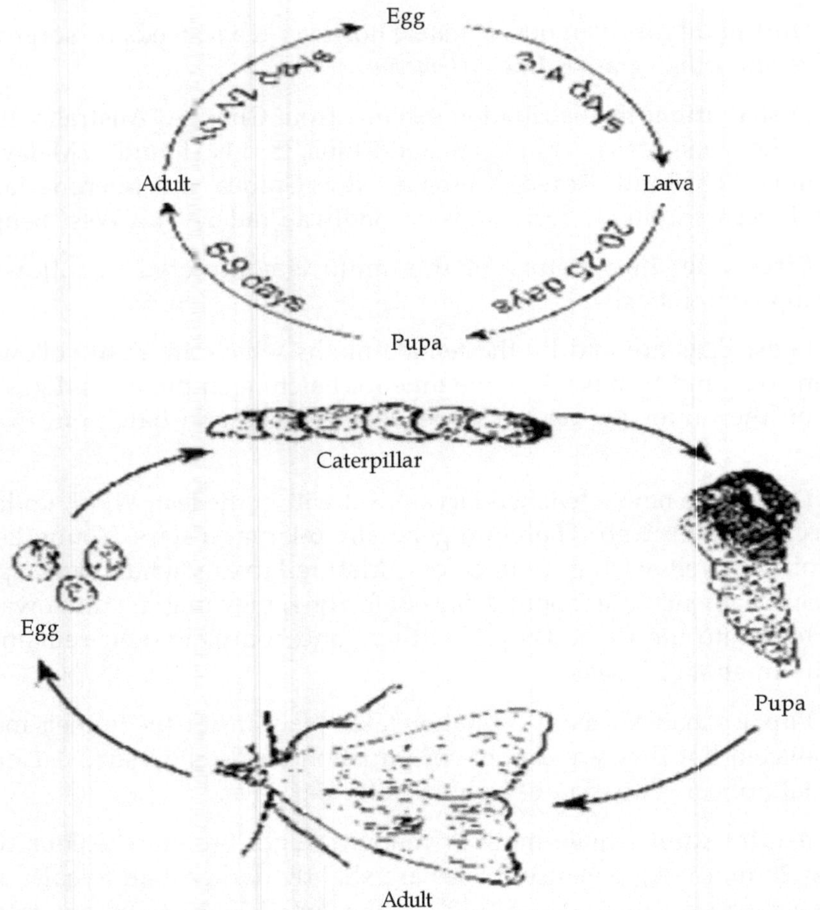

Life-cycle of *Ostrinia furnacalis*

Pests of Wheat

India has the largest area in the world under wheat cultivation. However due to low productivity, it's only the third largest producer of it after EU-27 and China. India's annual production of wheat has been around 75-79 million tonnes from 2006-2007 with production in 2008-09 estimated to be around 78.6 million tonnes. Wheat forms about 33% of India's total food grain production. Uttar Pradesh, Punjab, Haryana, Rajasthan and Madhya Pradesh are the main wheat producing states of India. Though this crop is less susceptible to pests however in recent years about half a dozen pests have come under consideration. Some important among them are discussed below:

1. **Pink Stem Borer**

Systematic Position

Order	:	Lepidoptera
Family	:	Noctuidae
Genus	:	Sesamia
Species	:	*inference*

Host plant: It affects wheat, sugarcane, maize, jowar rice, barley, oats and many species of grasses.

Distribution: It is found in many rice producing countries like China, Pakistan, India, Nepal, Bangladesh, Japan, Korea, Phillipines, Thailand, Sri Lanka, Vietnam etc. In India its found in the wheat fields of Madhya Pradesh, Uttar Pradesh, Haryana, Rajasthan etc.

Life-Cycle

Egg: Female moth lays eggs mostly in rows between the leaf sheath and stem. These eggs are bead like and yellowish in colour and incubation period lasts for a weak.

Larva: The larvas hatched are pink to purplish pink in colour. These are lighter on the ventral side. Entire body is smooth and cylindrical with a reddish brown head and about 26 mm long. They bore into the stem and make it hollow and attack several tillers before becoming full grown in 25-35 days.

Pupae Stage: Pupation normally takes place inside the stem and sometimes outside in hidden places such as between leaf sheath and stem. There is no cocoon formation and pupa is 18 mm long, dark brown with purple tinge on the head region. Pupal stage lasts for 5-10 days in summer and 15-30 days in winter.

Adult: Adult moths are stout fast flying insects having a wing span of about 4 cm. They are straw coloured with hairy body with forewing having a dark brown steak in the middle of forewing. The head and thorax bear tufts of thick brown hair.

Damage: The damage is caused by the caterpillars which bore into the stem after hatching and cause death of the cereal shoot known as "dead hearts" (Boring into the stem and making it hollow). It is most prevalent in the dry season just before the onset of monsoon. Infection of this pest considerably decreases the yield.

Control or Management

i) Removal of dead hearts and destruction of larvae restricts the pest control

ii) The stubbles should be collected and destroyed, when harvesting is over.

iii) Ploughing the field after harvesting crop kills dormant stages.

iv) Use of light traps should be made to catch the moths.

v) Spray of insecticides, such as carbaryl 0.02%, diazinon 0.02% at fortnightly intervals help to contain the pest.

vi) Application of 4% granules of diazinon, sevidolor endosulfan in root zone is more effective and doesn't kill the natural enemies.

vii) Use of Resistant varieties like Kalyan Sona could prove helpful and should be employed in susceptible areas

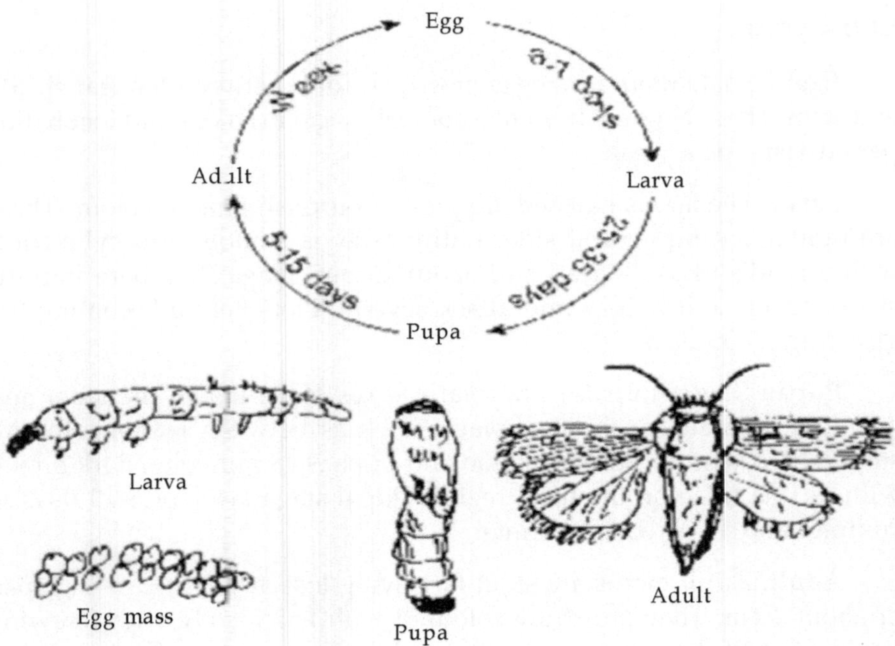

Life-cycle of *Sesamia inference*

2. **The Wheat Thrips**

Systematic position

Order	:	Thysanoptera
Family	:	Thripidae
Genus	:	Analpothrips
Species	:	*flavicinctus*

Host plants: Wheat, sugarcane and other such closely related plants.

Distribution: Recently it has gained the status of wheat pest and is found in various wheat cultivating states like Gujarat, Maharashtra, Uttar Pradesh, Punjab etc. It is important to note that low temperature favours pest incidence at early stages of the crop.

Life-Cycle

Egg: The eggs laid by the female moth are cotyledon or kidney shaped on the tissues of the tender leaves. Incubation period lasts for 6-12 days. These eggs hatch out within the leaf sheath.

Nymph: There are two nymphal instars which are completed in a time period of 2-8 days. This is followed by prepupal period that lasts for 1-3 days which in turn is followed by pseudopupal or pupal period that lasts for 1-5 days.

Adults: Adults moths are about 1.2 mm with fringed wings with sucking type of mouth parts. They are smooth, delicate and blackish in colour. These adults are ready to start the next generation very soon after emergence. The life-cycle is completed under favourable condition in almost 28-32 days producing many generations in a year.

Damage: Due to the presence of rasping-sucking mouth parts and also due to their invasion of tender leaf tissues they suck the plant sap easily and voraciously, that results in curling of leaves – producing wilting type of symptoms. The affected plant show stunted growth with characteristic brown patches on the leaves. In chronic cases the white streaks develop on the leaves.

Control or Management

i) Destruction of stubble after harvesting.
ii) Destruction of eggs and larvae.
iii) Use of Dust 5% BHC.
iv) Spraying 0.02% phosphamidon or 0.03% diazinon reduces the moth population upto a large extent.
v) Removing the leaf sheaths.

3. **Lesser Grain Borer**

Systematic Position

Order	:	Coleoptera
Family	:	Bostrychidae
Genus	:	Rhyzopertha
Species	:	*dominica*

Host plants: Wheat, maize, rice and on stored grains etc.

Local names: American wheat weevil, Australian wheat weevil, grain weevil, stored grain eater. In India its locally called lal surhi, thuthun heen surhi.

Distribution: It is a serious pest of dried stored products throughout the tropics and is also found in temperate countries. Though its globally found but it is serious pest of Indian, Pakistani, American and Australian wheat fields. It is believed that this pest has spread due to international trade of its subjected products.

Life-Cycle

Eggs: Eggs are laid by the beetle after 4 days of emergence. These eggs are voidal in shape, 0.6 mm in length and 0.2 mm in diameter laid loosely in grains. When initially laid they are white and then turn brown just before hatching. A female is capable of laying 300- 600 eggs in her life time. The hatching period lasts for 5-15 days

Larva: The newly hatched larvae are white to cream coloured with typical bitting type of mouth parts and three pair of legs. They are active and soon after emergence they make their way into the grain and complete their rest of period there. The larvae are quite mobile and linear in shape at first but become almost immobile and crecent shaped as they develope. The grub period is last for 35-45 days.

Pupa: Pupation generally takes place inside the grain but sometimes larvae feeding on the powdered starchy material outside the grains undergo pupation there. Pupa is usually dull white in colour with irregular body shape. This period lasts for 6-8 days

Adults: The adult is 2-3 mm long reddish brown in colour with a slim cylindrical shape. The elytra (wing cases) which cover the membraneous hind wings, have regular rows of coarse punctures (finer at sides) covered with curved, setae (Hair). The front edge of the pronotum has a saw toothed appearance. The antennae are 10 segmented and terminate in a prominent club. There is a prominent constriction between prothorax and elytra.

The antennae are 10 segmented and terminate in a prominent club. There is a prominent constriction between prothorax and elytra.The head is not visible when viewed from above. Adults are capable to live for 240 days, emerge by chewing through the outer grain layers.

Damage: Both larva (grub) and adult cause the damage by boring or eating grains. The grubs usually eat out the starchy contents of the grains leaving the outer husk only. The adults destroy the whole grain by chopping its husk as well.

Control or Management

i) Removal of adult insects from the grain by sieving can reduce population.

ii) The addition of dusts such as ash and clay can kill insects by causing the insects to die from desiceation.

iii) Removing split grain from around storage facilities can reduce lesser grain borer populations.

iv) The use of fungus–*Beauveria bassiana* can be used as a biological insecticide.

v) Storage places or graneries should posses low oxygen concentration and equally high CO_2 content, proves helpfull for controling this pest.

vi) Use of fumigant phosphine can be used in sealed storage facilities

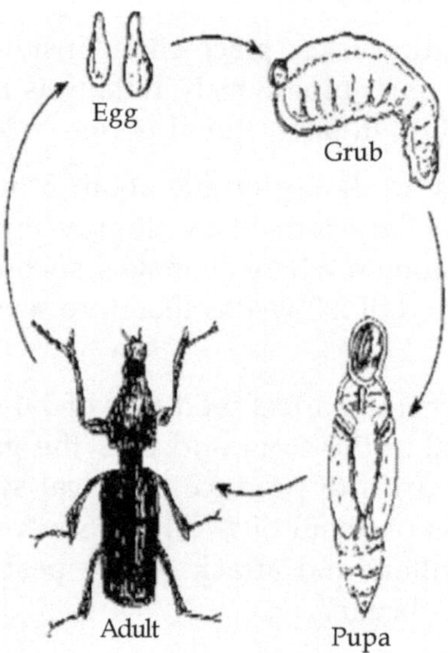

Life-cycle of *Rhyzopertha dominica*

1. **Sorghum shoot fly Systematic** Position:

Order	:	Diptera
Family	:	Anthomyidae
Genus	:	Antherigona
Species	:	*soccata*

Host plants: Sorghum, maize, wheat, various grasses and wild plants.

Distribution: It is widely distributed in Europe, Africa and Asia. In India its found mostly in Maharashtra, Madhya Pradesh, Arunachal Pradesh, Karnataka and Tamil Nadu.

Life-Cycle

Life-cycle involves following four developmental stages.

Egg. The female lays white cylindrical, distal somewhat flattened 40 eggs singly on the seedlings mostly on the basal half of lower surface of the centrally located leaves and also on the tillers. These eggs are provided with two wings like lateral projections. Incubation period lasts for 2-4 days.

Larva: The larval period is of 10-13 days. The newly hatched larva is dirty whitish in colour and later becomes yellowish. These (maggots) are cylindrical, tappering towards head, about 8 mm in length. They reach in between the sheath and the axis and finally bore into the shoot where they feed on decaying tissue.

Pupa: The population takes place either inside the stem (bored) or in the soil. Pupa is of oblet type which initially is light brown in colour and changes to deep brown later. Pupal period = 5-6 days.

Adult: The adults are housefly like about 5 mm long of dark grey colour. The female is characterised by the presence of 6 (3 pair) dark spots on pale grey abdomen where as male 4 such dark spots.Since, one generation takes a period of 2-3 weeks therefore several such generations occur within a year

Damage: Pests generally affect seedlings of 3-4 weeks age and tillers. The maggots bore inside the stem and cuts the growing point due to boreing central shoot dries to produce a typical symptom called 'dead heart'. They cause loss of about 60% during severe attack. The infested plant produces side tillers and attack of this pest flourishes highly in between October to February.

Control or Management

i) Removal crop escapefrom the pest.

ii) Use of high seed

iii) Use of disease resistant varieties like Co 18, CSH–1, CSH–15 may be grown to reduce the damage.

iv) Use of seeds pelleted with insecticides.

v) Ploughing soon after harvesting and destruction of stubble could prove beneficial for pest control.

vi) Seed treatment with imidaclo-prid 70 Ws @ 10 kg/1 kg of seeds.

vii) Spraying of 0.025% methyl demeton is also effective rate 13 kg/ha and destruction of infested seedlings after 3-4 weeks of sowing.

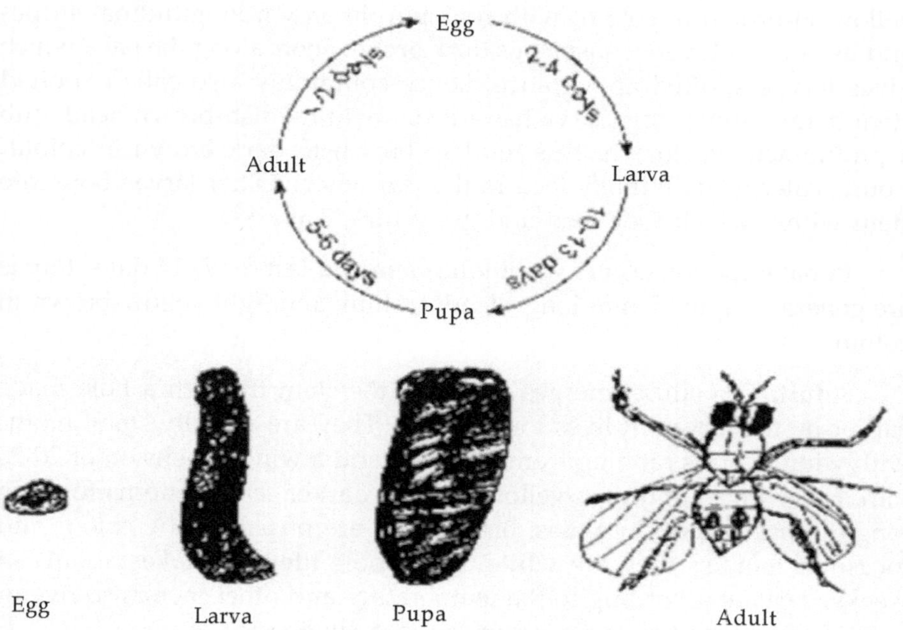

Life-cycle of *Atherigona soccata*

2. **Sorghum Stem Borer**

Systematic position

Order	:	Lepidoptera
Family	:	Crambidae
Genus	:	Chilo
Species	:	*partellus*

Host plants: Sorghum maize, sugarcane, paddy, bajra and several other grasses.

Distribution: It is a native of Asia and also found in Eastern, Western and Southern Africa. In India its found in Uttar Pradesh, Karnataka, Madhya Pradesh, Arunachal Pradesh, Maharashtra etc.

Life-Cycle

Egg: The female lays flat and oval-scale like creamy-white about 0.80 mm long eggs in clusters on the upper and underside of leaf surfaces mainly near the mid ribs. They hatch after 4-10 days after laying.

Larva: There are six larval instars, larvae are creamy white to yellowish-brown in colour, with four purple-brown longitudinal stripes and usually with very conspicous dark brown spots along the back which gives larvae spotted appearance, hence commonly also called spotted stem borer. Fully grown larva has a prominent reddish-brown head with a prothoracic shield which is reddish brown to dark brown in colour. Young caterpillars initially feed in the leaf whorl. Older larvae bore into stem within which they feed and grow for 2-3 weeks.

Pupa: Pupation occurs within the stem and last for 7- 14 days. Pupae are generally upto 15 mm long, slender, shiny and light yellow-brown in colour.

Adult: The adult emerges out from the stem through a hole made earlier by the larvae. It lives for 2-6 days. They are actually small moths with wing lengths ranging from 7-15 mm and a wing expension of 20-25 mm. Forewings are brown-yellowish with darker scale patterns forming longitudinal stripes. In males, hind wings are of pale straw colour and incase of females they are white. The whole life-cycle takes about 3-4 weeks, varying according to the temperature and other factors so five or more generations are easily produced in a single year.

Damage: It generally attacks the older plants. Damage occurs as a series of small holes in lines in younger leaves on patches of transparent leaf epidermis in older leaves. Since the caterpillers are the internal tissue feeders are causing death of central shoot – "dead hearts" causing breakage of stem or drying of plant that ultimately leads to the death. It causes damage upto 80% when condition favouring pest persists.

Pest management or control

i) Removal and destruction of dead hearts.

ii) Burning of stubbles after harvesting.

iii) Intercropping with non host crops like cowpea.

iv) Use of pheramone traps for collecting the pests.

v) Application of granules or dusts to the leaf whorls early in the crop growth to kill early larval instars.

vi) Neem products are found very effective and may be applied to the leaf whorl in a 1 : 1 mixture with saw dust or dry clay.

vii) Introduction of Trichograma, Telenomus as the egg parasites could restrict the pest population upto huge extent.

3. **Sorghum Shoot Bug Systematic** Position

Order	:	Hemiptera
Family	:	Delphacidae
Genus	:	Peregrinus
Species	:	*maidis*

Host plants: This pest mostly effects sorghum and maize sometimes sugarcane is also suggested as its host along with oat and some other species of grasses.

Distribution: It is native of Africa and is found throughout the World in Tropical areas such as Bermuda, Caribbear, Cuba and Phillipines.

Life-cycle

Egg: The female bug makes slits in the midrib of the leaves with its ovipositors and lay 2- 4 eggs per oviposition site. The female deposits a drop of glue like material after oviposition sealing the eggs. About 100 white, enlongated and cylindrical eggs are laid by the female during the oviposition period of 7 days-with an average rate of 15 egg are per day.

Nymph: There are about 4-6 instars involved. Nymphs are initially whittish in colour and become darker and yellowish in later instars. The wing pads appear during 4th instar and they initially feed within the covered areas like leaf sheaths, leaf whorls etc. As they grow they spread out to wider areas. They take about 15-20 days to become adults depending on the temperature and nutrition.

Adults: The adult insect is about 3.5 mm long with transluecent wings and yellowish brown in colour. Several overlapping generations occur in a year because a complete generation involves almost 30 days.

Damage: The nymphs and adults both cause damage to the host plants. They cause injury to the plant, sucking of plant sap, tissue damage resulting from oviposition. Moreover they secreate honey-dew which supports the growth of sooty mold which impedes photosynthesis in severe conditions. Both adults and Nymphs are capable of virus transmission, which results in yellowing and stunting of the plant – like corn mosaic in maize and 'freckled yellow' in sorghum.

Pest Control and Management

i) Crop rotation limits damage.

ii) Field margins should be focused.

iii) Mass collection and destruction of insects by pheramone and light traps

iv) Introduction of larval and egg parasites could restrict insect population.

v) Use of suitable pesticides like spraying 625 ml of melathion 50 EC in 500 ltr. of water per hectare.

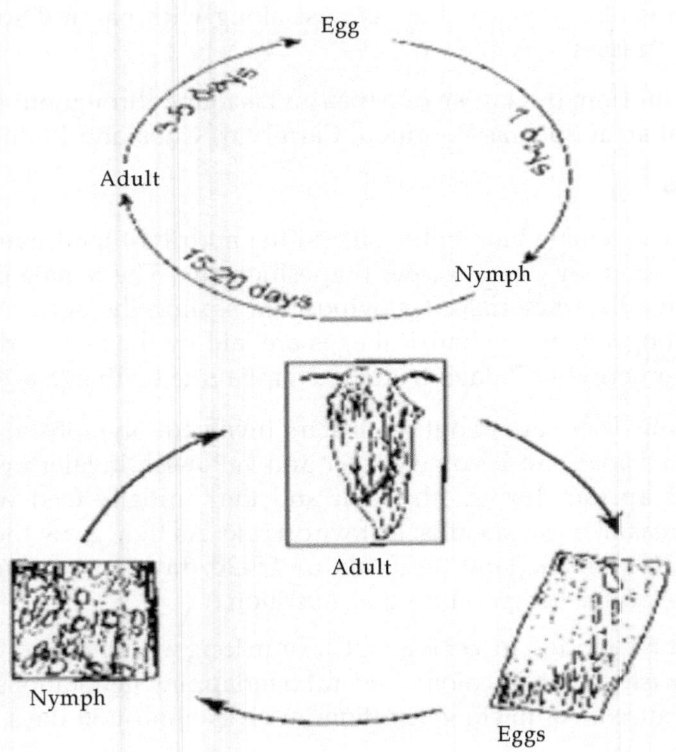

Life-cycle of sorghum shoot bug

Pests of Sugarcane

India is the largest sugarcane producer in the world only after Brazil with its annual production of 28,50,29,000 tonnes. Since this crop is a long duration crop of 10-18 months and is therefore liable to be attacked by the pests. More than 200 insect species have been identified that cause damage to this crop. Some of the important among them are:

1. **Sugarcane Root Borer**

Systematic position

Order	:	Lepidoptera
Family	:	Pyralididae
Genus	:	Emmalocera
Species	:	*depresella*

Host plants: Sugarcane is its main host, though it could also infest maize, sorghum etc.

Distribution: It is found in almost all sugar growing areas of India but it is a serious pest of Eastern Indian states like Uttar Pradesh, Punjab, Madhya Pradesh, Bihar etc. At lower scale its also found in Pakistan Punjab etc.

Life-Cycle

Egg: The female lays about 250-350 eggs after coupulation in the night hours either singly or in masses on the lower surface of leaves, the stem or on the ground. These eggs are oval and creamy white in appearance. The eggs hatch out in 6-8 days.

Larva: Soon after the emergence of first instar larva it bores into the stem near the roots and feeds inside. This larva is about 2.0 mm long of ligh yellow or white colour with brown head. As they feed they cut across the stem, reaching the adjoining tillers. The larval period is about 35-45 day during which the larva attains a maximum growth of 25-30 mm and before changing into pupa the larva move above the soil surface within the stem and makes a small hole for its exit and constructs a silk tube for further development.

Pupa: Pupation period lasts for about 15 days and occurs within the pre constructed silken tube, pupa is reddish brown in colour about 20 mm long. They are last instar larva under goes *dipause* to cross winter.

Adult: The adult moths have pale-pink head and have white hind wings which are somewhat larger in width than forewings. Body measures about 30 mm long with an average wind expanse of 32 mm. The abdominal tip of male is tappering while that of female is cylindrical.

Damage: From the month of April to June during which the crop is in early stage the caterpillars feed inside the lower stem producing dead-heart, due to which the plants die within 4-8 weeks. The infected plants which survive despite being attacked produce very little yield. Though the dead hearts caused by the root borer resemble with those of stem borer but they are quite differentiable as there is the single hole present on the stem of infected plant in case of root-borer where in case of former they are more than one in number.

Control or Management

i) Proper and deep ploughing of field reduces pest population.

ii) Burning the stubbles in the field and ratooning should be avoided in endemic areas.

iii) Cutting of infested shoots at ground level.

iv) Biological control *via* trichogramma minutum (egg parasite), stenobracon deesae (larval parasite) etc.

v) Treatment of soil with Endosulfan @30 kg/ha. Has been found useful in controlling this pest.

vi) Resistants varieties like CO 313, 513, 1158 etc. should be introduced in the damaged areas.

vii) At the time of planting dusting of 10% BHC @ 30-35 kg per hectare was also found effective.

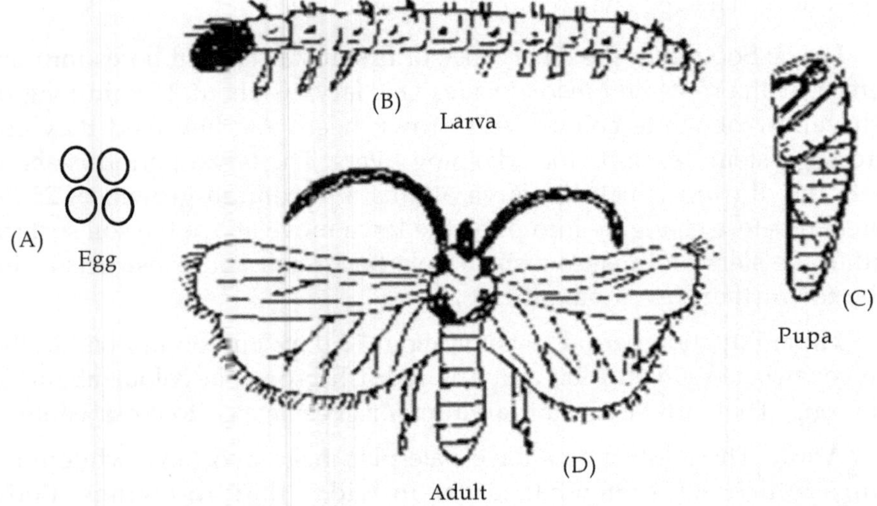

Life-cycle of *Emmalocera depressela*

2. **Sugarcane Stem Borer**

Systematic position

Order	:	Lepidoptera
Family	:	Pyralididae
Genus	:	Chilo
Species	:	*infuscatellus*

Host Plants: A part from sugarcane, it feeds on maize, jowar, bajra etc.

Distribution: It is considered one among the serious pests in India, Pakistan and Afghanistan. It is also reported from Burma, Indonesia, Taiwan, Japan, Korea and Phillipines. In India it is widely distributed in Uttar Pradesh, Madhya Pradesh, and Punjab etc.

Life-Cycle

Egg: The female lay scale like creamy-white egg masses on underside of leaves 3-6 rows on both sides of the midrib. Each mass consists about 30 eggs which hatch out in 5-7 days and there are about 300-400 eggs lay by a female in her life time within a time period of 2-4 days.

Larva: Five remarkable instars are found in a time period of 3-5 weeks. The newly hatched larva feed on tender leaf sheath for a week period later it makes a hold on the stem just at the soil surface where it feeds on plant tissue for next 3-4 weeks and then undergoes pupation. Initially larva has a dark head with translucent body but in later instar it is dull white with brownish-red coloured longitudinal stripes on the trunk.

Pupa: Pupation occurs inside the larval tunnel in the stem and lasts for 6-10 days. The pupa is brownish in colour and comes out through the exit hole that has been made by the larva just before pupation, which represents the adult moth. It should be noted that pupation does not occur in winter and its the last instar larva which undergoes diapause to escape winter thereafter in month of March it (pupation) occurs.

Adult: The adult moths have straw coloured fore- wings and white hind wings. The wing expension is about 3.5 cm in females and 2.5 cm in males. The posterior region of the abdomen is somewhat pointed. The adults survives for 8 days during which they copulate in night and just after 2-3 days of copulation female starts laying eggs. About 5-8 generations occur annually.

Damage: Since the first instar larvae feed on the leaves hence they damage leaves, later the 2nd, 3rd and 4th generation caterpillars damage the crop by consuming internal tissue of the stem where they have bored. The attacked plants show 'dead heart' and dry up completely in later stages. The larva bored move from the ground level stem towards the upward direction internally producing tunners, but never reach to the top. The presence of this pest inside the plant is indicated by an 'offensive smell' on removal of dead hearts. The long tunnelling of stem reduces the yield upto 30%.

Control or Management

i) Primary measures include removal and destruction of 'dead hearts'.

ii) Burning of stubbles after harvest.

iii) Early planting of crop to escape the pest has been found effective in north indian states.

iv) Collection of moths by light or pheramone traps.

v) Spraying with 0.5% Endosulfans 5% malthion or 2% parathion at 3 weeks interval is also found effective in controlling the pest.

vi) Treating soil with 2% lindane dust @ 30 kg/hectare before sowing is also effective.

vii) Biological controls include egg parasites like *Trichogramma minutum* and larval parasites like Bracon

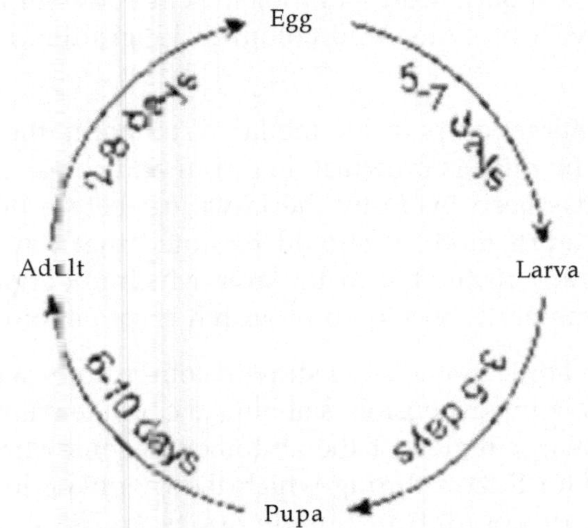

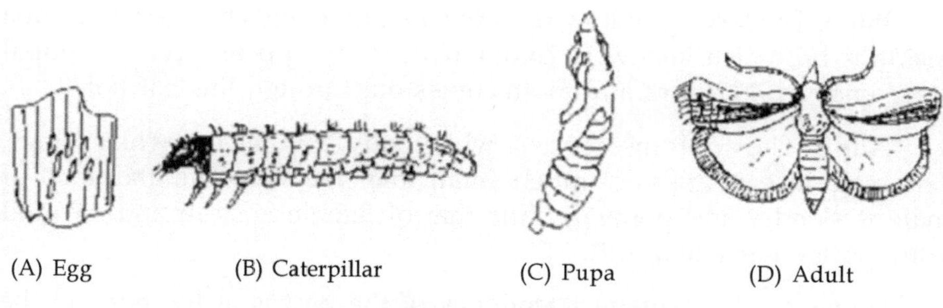

(A) Egg (B) Caterpillar (C) Pupa (D) Adult

Life-cycle of *Chilo infuscatellus*

3. **Sugarcane top borer**

Systematic Position

Order	:	Lepidoptera
Family	:	Pyralididae
Genus	:	Scirpophaga (Trporyza)
Species	:	*novella*

Host plants: Apart from sugarcane, its also found on Jawar, bajra, sarkanda, Dhab and other grasses.

Distribution: This pest is distributed throughout South Asia countries where sugarcane is cultivated like India, Pakistan, Thailand, Burma, China, Phillipines etc. In India its specially found Uttar Pradesh, Madhya Pradesh, Bihar, Punjab, Tamil Nadu and Mysore.

Life-cycle

Four development stages are found in its life-cycle they are:

Egg: The female lays 300-350 eggs either singly or in groups of 5-10 on the under side of the leaves. These egg groups are covered with a brown tuft of hairs and are quite prominent. Incubation period is of 5-8 days. The eggs laid are enlongated; oval shaped which overlap one another in the cluster.

Larva: The eggs hatch into larvae which are about 2 mm in size with black head and these young larvae bore the top shoot of the cane through the mid rib of the leaf occupied. The larval stage lasts for 35-45 days during which five stages or instars of larval development occur. The fully grown larva is about 30 mm long and yellowish in colour that forms a characteristic chamber with an emergence exit just above the node. During larval progression the larva contineously grow in size and go on penetrating the top till 4-7 internodes are bored.

Pupa: The larva pupate within the already made chamber and pupa inside is brownish yellow in colour with shiny appearance. The pupal stage ends in 7-12 days and moth comes out through the exit hole.

Adult: The adult insect has a white body with silvery white wings. The male insects are very much small than females. The abdomen of male is slender and pointed while that of female is stout and covered with orange hairy anal tuft.

Damage: The damaging tendency of the pest is at the peak in the months between Marchs to September. The first two instar larvae attack young plants before the formation of canes, which usually die causing total demage to the yield. These larvae make holes in the midrib of leaves and then travel in the central shoot and consume the growing plant tissue of the top part of the plant. The central shoot tunneled by the later instar larvae dries up and resulting in the formation of 'dead-hearts'. The formation of side shoots after the damage of central shoot. Which gives rise to a bunchy top is an striking symptom of this pest. Almost 25-30 of loss is caused by the caterpillars of this pest.

Control or Management

i) The egg masses should be collected and destroyed.

ii) Moths should be collected and destroyed using light traps.

iii) Attacked shoots should be cut at the ground level.

iv) Ratooning should be avoided where pest attack is serious.

v) Biological control includes, *Trichogramme chinolis* (Egg parasite), *Xanthopimla mursei* (Larval parasite), *Stenobracon deesae* (pupal parasite) etc.

vi) Growing resistant varieties of crop like Co 419, CoS 767, Co 313, 453, B.O 10, 24 etc.

vii) Soil application of carbofuran at 2 kg a.i./hectare is recommended.

viii) Moreover spraying the crop with 0.1% endposulfan (35 EC) @ 1200-1500 litres/ hectare has minimized the attack.

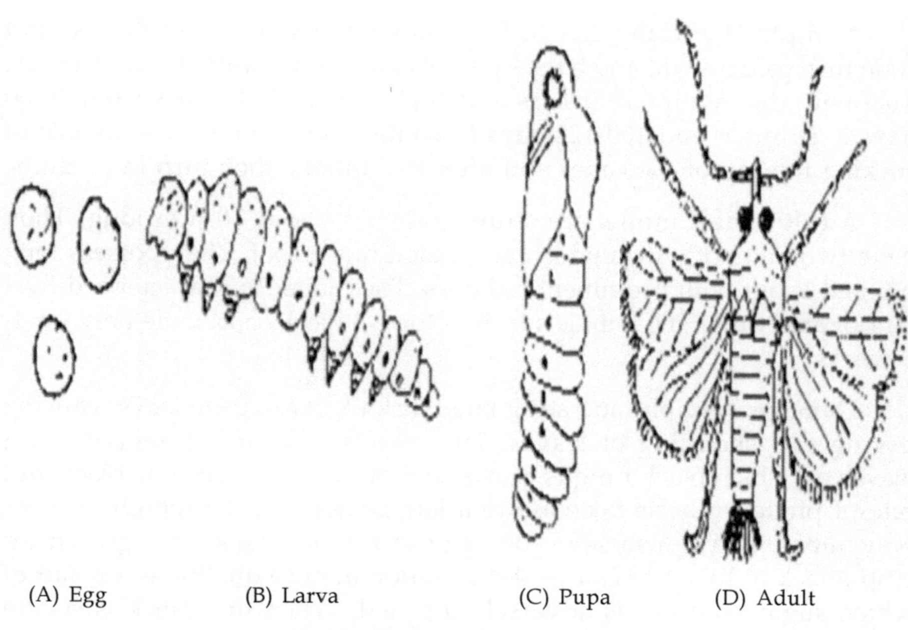

Life-cycle of *Trporyza novella*

4. **Sugarcane Leaf Hopper**

Systematic Position

Order	:	Hemiptera
Family	:	Fulgoridae/Lophopidea
Genus	:	Pyrilla
Species	:	*perpusilla*

Host plants: Important hosts include sugarcane, bajra, jawar and Maize.

Distribution: It is found in India, Sri Lanka, Myanmar and Thailand. In India its found in almost all sugarcane growing states like Uttar Pradesh, Bihar, Punjab, Madhya Pradesh and Maharashtra.

Life-cycle

There are following three stages in its life cycle, *i.e.,*

Egg: The female lays eggs in large clusters of pale greenish yellow each containing 10-60 eggs on the lower surface of the leaves and are covered with cottony waxy filaments, secreted by anal tufts. A single female is capable of laying 750 eggs in one generation. Eggs laid are oval and shiny and hatch out in 8-10 days in summer and 20-40 days in winter.

Nymph: The freshly hatched nymphs are cream white coloured and soon turn pale brown, and have a pair of anal waxy filaments like brushes. There are five nymphal, instars and Nymphal period varies from 40-60 days in summer and 100-120 days in winter. The nymphs are capable of sucking the sap of the canes and after five moults they turn into adults.

Adult: Adult moths are straw coloured about 10 mm long. They have two pairs of wings which are folded like a roof. They possess long pointed snout with prominent red eyes. The female is characterised by a pair of anal tufts. The female survives for 6-8 weeks and male only for 4-6 weeks.

Damage: Nymphs and adult bugs suck plant sap from leaves causing drying and shedding of leaves. The insects discharge honey-dew on leaves on which black fungus grows so that the leaves become black and rate of photosynthesis decrease to a large extent. In the month of July–November, they cause sever demage to ratoon crops. In a grown up crop attack of this pest causes deterioration in juice quality as a result of which sugar recovery is adversely affected. When the attack is severe cane looses about 50% of sucrose.

Control

i) Collection and destruction of egg masses.

ii) Burning all trash after harvesting.

iii) Bagging of adults in nets.

iv) Moderate use of Nitrogen fertilizers.

v) Stripping of dried leaves.

vi) Ratooning should be avoided in severely.

vii) Resistant varieties should be used like Co 386, Co 975 etc.

viii) Spraying of dimecron, dimethoate, carbaryl etc.

ix) Biological control includes *Tetrastychus pyrillae* (Egg parasite) *Dryinus pyrillae* (Nymph parasite) etc.

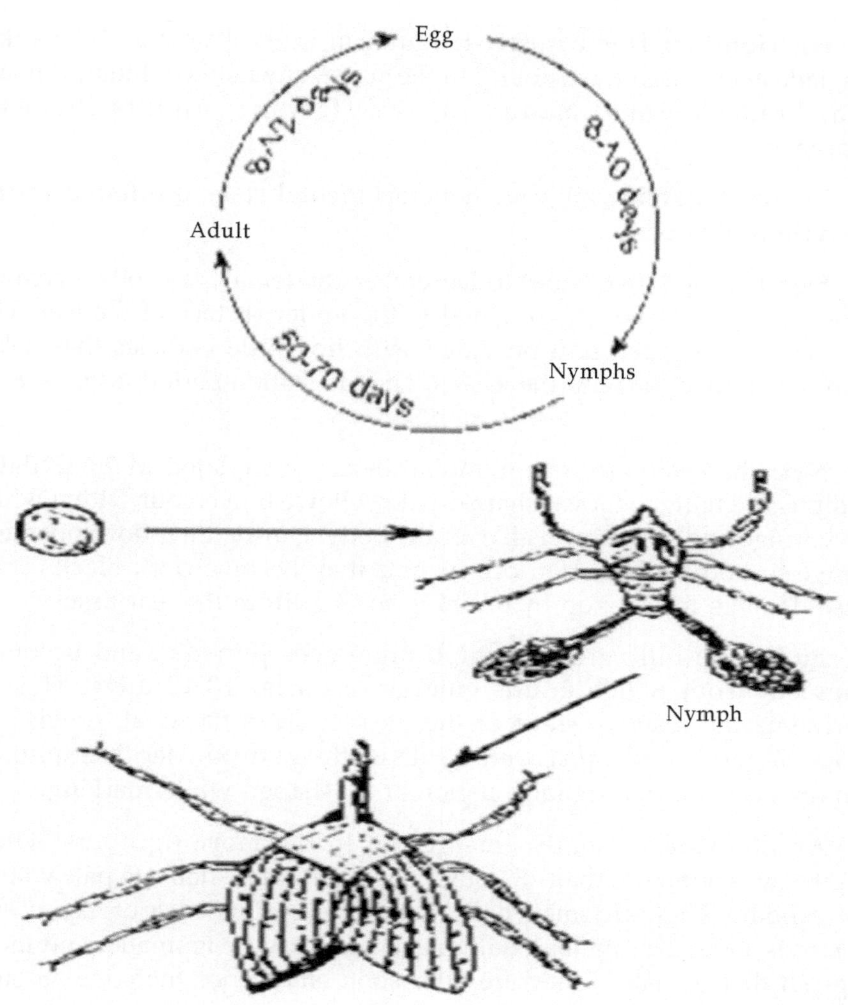

Life-cycle of *Pyrilla perpusilla*

5. **Sugarcane Whitefly**

Systematic position

Order	:	Hemiptera
Family	:	Aleyrodidae
Genus	:	Aleurolobus
Species	:	*barodensis*

Host plants: The main host is sugarcane; however it also feeds on jawar, bajra, maize, wheat, barley etc.

Distribution: This whitefly is found in India, Pakistan, Sri Lanka, Bangladesh etc. In India its found in the sugarcane fields of Bihar, Punjab, Uttar Pradesh, Tamil Nadu, Gujarat, Haryana, Andhra Pradesh, Uttaranchal etc.

Life-cycle: Following four developemental stage are found in the life-cycle of this pest.

Egg: During Novemeber to December the female lays 60-65 creamy white conical eggs which are glued to the under surface of the leaf. The eggs are cone shaped and provided with tiny little pedicles that helps them to get attached with the leaves. The incubation period lasts for 8-10 days.

Nymph: There are four nymphal instars completed in 26-32 days. Nymphs are flattened, oval shaped pale yellowish in colour with a white waxy fringe and white meal all over the body. During their developement change in body colouration occurs and they become dark black. They cause damage to the crop by draining the sap from the leaf tissues.

Pupa: The fully grown nymph undergoes pupation and becomes quiescent from which adults emerge out after 10-12 days. Thus it represents the inactive stage of the pest. Pupa is flat oval, greyish in colour, larger than nymph covered with white waxy powder. The splitting of pupa occurs at the thoracic region at 'T' shaped white marking.

Adult: Adults usually emerge out in the morning hours. Their longitivity is not more than 48 hours. These are small, delicate pale yellow Life 48 hours. These are small, delicate pale yellow Life-cycle of *Aleurolobus barodensis* about 2-4 mm long with mealy wings. Male is smaller and more sluggish than female. They are after soon emergence the copulate and eggs are laid within half an hour.

Damage: Nymphs and Adults, both are capable of damaging the crop but nymphs are more destructive and can be considered as, sole destroyers of crop. They damage the crop by sucking the cell sap so that yellow streaks appear on the attacked leaves and the crop acquires a paleish green appearance in the months of July–November. They cause severe damage to the ratoon crops. Incase of severe infestation, leaves turn black due to the deposition of fungus on the honey dew secreted by the pest. Sugar recovery is reduced by about 20-30%.

Control

i) Removal of infested leaves.

ii) Ratooning of the crop should be avoided.

iii) Use of resistant varieties likes Co 285, CoS 1158 etc. minimise the pest attack.

iv) Spraying the crop with endosulfan 35 EC–1.5 liters is also recommended.

v) Other chemical measures include use of insecticides like dimecron, Nuvacron, LVC melathion etc.

vi) Biological control includes *Cardiogaster secundus* (Egg Parasite), *Azotus delhiensis* (Nymph parasite), *Amitus aelurolobi* (Pupal parasite) etc

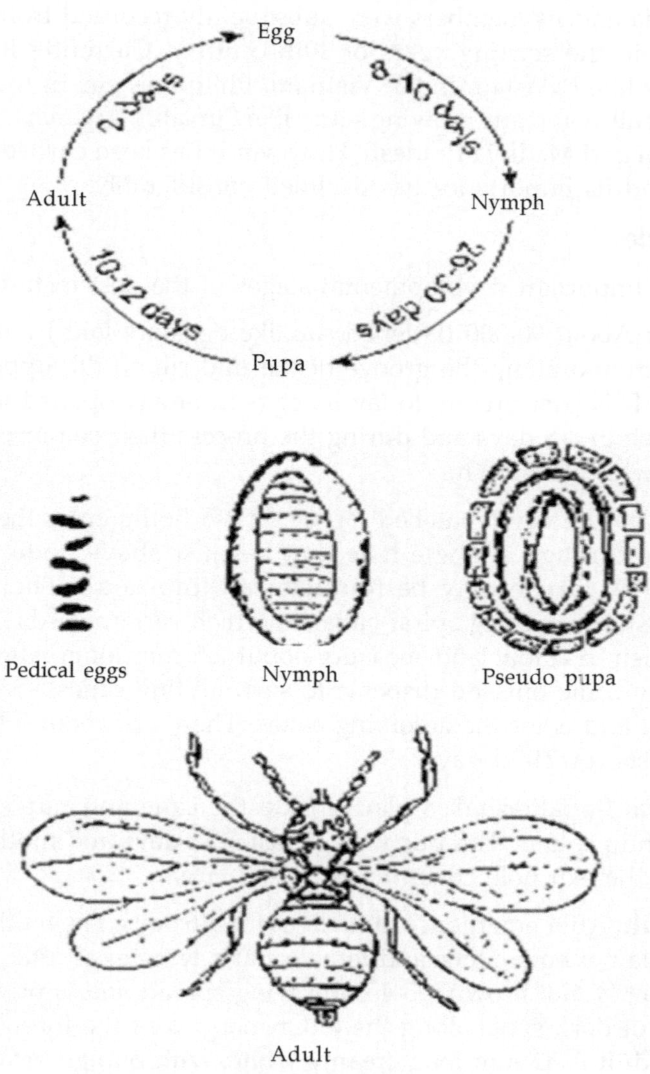

Life-cycle of *Aleurolobus barodensis*

5. **Gurdaspur Borer**

Systematic Position

Order	:	Lepidoptera
Family	:	Pyralididae
Genus	:	Acigona
Species	:	*steniella*

Host plants: Sugarcane, Maize, Jawar, etc.

Distribution: This pest was first discovered from Sialkot (Pakistan) then India and its members were subsquently recorded from Gurdaspur (Punjab) in the starting years of 20th century. Currently it is found in countries like Pakistan, India, Vietnam, Philipines etc. In India its found in almost all sugarcane growing states like Punjab, Haryana, Uttar Pradesh, Rajasthan and Madhya Pradesh. However it has been curbed upto a large extent and its importance has declined considerably.

Life-cycle

The important developmental stages of the pest include.

Egg: About 90-300 flattened sac like eggs are laid by the female in clusters in and along the groove of the mid rib on the upper surface of the leaf. This pest prefers to lay its eggs in newly opened leaves. These eggs hatch in 3-8 days and during the process their colour changes from creamy white to brown.

Larva: The newlyhatched larvae, in 4-5 hours enter the top portion of a cane through a single hole that lies just above node. During this stage 30-60 larvae may be found in an internode. There they feed voraciously by making spiral gallaries which run upwards. These larvae are pinkish in colour and measure about 2.5 mm long. After about 7-10 days they come out and disperse to surrounding canes, sometimes they came out and enter the adjoining canes. There are about 5 larval instars and it lasts for 21-35 days.

Pupa: Pupation takes place inside the cane and pupa produced is brownish in colour. The pupal stage last 6-12 days and moth emerge out through the exit hole constructed by the larva.

Adult: After emergence moth lives for 4-6 days. The moth's coupulate on 2nd day of emergence and female starts lying eggs after 2 days of it. The insect is dull brown, 25-45 mm wing spread and is provided with a number of dark spots along the outer margins of the forewings. A fully grown adult is 32 mm long creamy white, with orange brown head and white hind wings.

Damage: Soon after the on set of monsoon attack of this pest begins, in the initial stages the larvae feed vocariously in the top portion of the canes. The caterpillar after entering the internode moves upwards by making spiral gallaries as a result canes become weak, dry up or fall down easily in strong wind or heavy rain. The pest destroys 20-25% of the crop.

Control

i) The roughing should be done when young caterpillars feed gregariously.

ii) Do not ratoon heavily affected crop.

iii) The attacked shoots should be regularly removed and destroyed.

iv) Use of resistant varieties should be prefered like Co 6812, 1157 etc.

v) Plough up the fields not meant for ratooning and desctroy the stubble before monsoon.

vi) Biological control includes *Trichogramma* (Egg parasite), Apanteles flavipes (Larva, parasite), and isotima javensis (pupal parasite).

vii) Use of chemicals is not feasible. India was probably first country in the world that domesticated cotton for fabrics and ranks seconds only after China in production (33.00 million bales/year). Despite of the fact that India uses about 25% of total cotton land of globe yet it accounts for only 20% of world cotton production. The cause for low cotton productivity of india is due to following reasons:

   i) More than 65% of cotton-cultivated lands depend on rainfall and can't be irrigated delibrately.

   ii) Indian cotton fields are susceptible to a wide range of insect pests. More than 1300 species of insects have been reported World wide that attack cotton. In India only 162 species have been recorded, among which only 16 are considered as major pests.

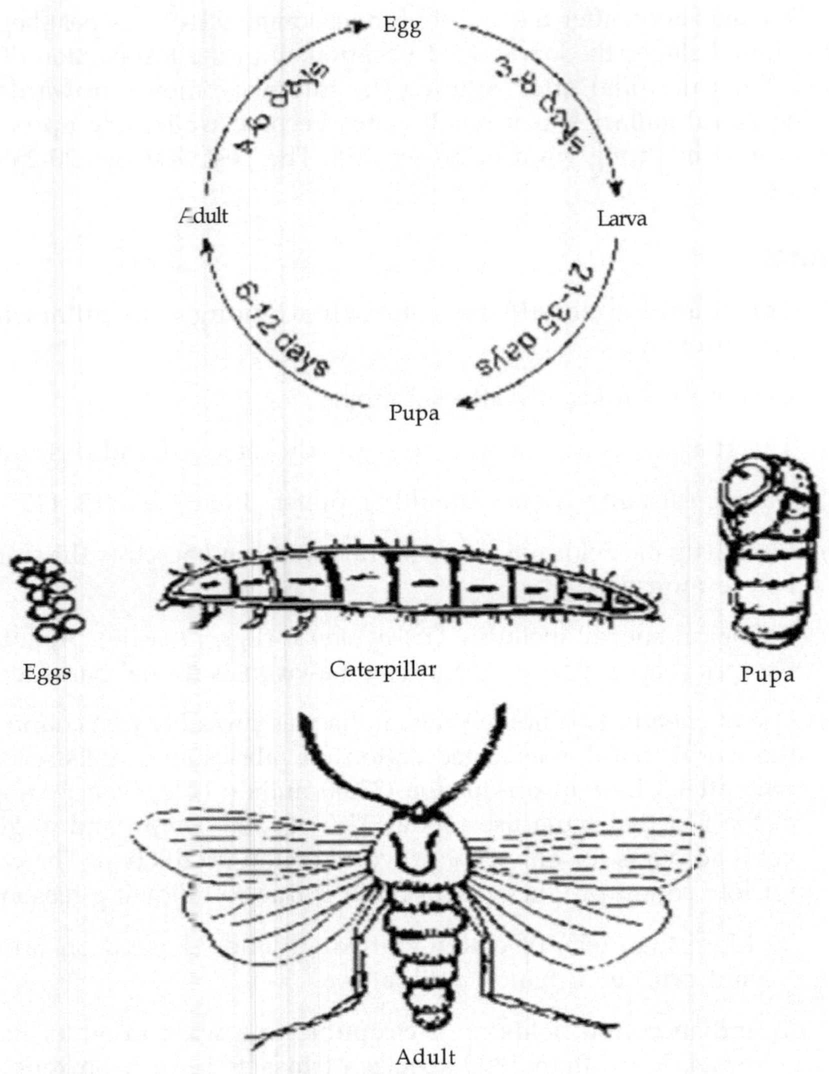

Life-cycle of *Acigona steniella*

Pests of Cotton

India was prcbably first country in the world that domesticated cotton for fabrics aнd ranks seconds only after China in production (33.00 million bales/year). Despite of the fact that India uses about 25% of total cotton land o: globe yet. It accounts for only 20% of world cotton production. The cause for low cotton productivity of India is due to following reasons:

i) More than 65% of cotton-cultivated lands depend on rainfall and can't be irrigated delibrately.

ii) Indian cotton fields are susceptible to a wide range of insect pests. More than 1300 species of insects have been reported World wide that attack cotton. In India only 162 species have been recorded, among which only 16 are considered as major pests. Some of the important among them are described below:

1. **Pink Boll Worm**

Systematic Position

Order	:	Lepidoptera
Family	:	Gelehiidae
Genus	:	Pectinophora
Species	:	*gossypiella*

Host plants: Cotton is the major host of this pest. Apart from this it also feeds on maize lady's finger, various members of families like malvaceae and tilaceae.

Distribution: The pink boll worm is native to Asia (India) but currently found in almost all cotton growing countries. In India it is found in all cotton growing states like Punjab, Gujarat, Maharashtra, Andhra Pradesh and Tamil Nadu.

Life-Cycle

Following four developmental stages are found in the life-cycle of the insect.

Egg: The female moth lays eggs either singly or in clusters (1-10) on the underside of the young leaves, new shoots, and flower buds and in the young green bolls. The eggs laid are flat and whitish in colour, which turn red just befor hatching. A female lays about 200 eggs on the plant which hatch in 2-10 days. However the hatching period may extend upto 30 days depending upon the environmental condition.

Larva: Larval period lasts for 10-30 days including 4 larval instars young larvae are white or colourless about 1mm long and enter the flower buds, the flowers or the bolls. Inside the boll they consume seeds of the plant most importantly the entry hole gets quickly healed up and it is difficult to distinguish between infested and normal uninfested boll. The fully grown larva with characteristic brown head is about 1 cm long, white in colour with a double red band on upper part of each segment.

Pupa: The fully grown larvae of about two weeks come out of the holes for pupating on the ground, among fallen leaves, debris etc. It should be noted that pupation takes place inside silken cocoon and pupal period is of 6-20 days. Initially pupa is light brown in colour which later changes to dark brown, its about 6 mm long and 2.5 mm in width.

Adult: The adult moths are dark brown in colour, with many black spots on forewings and fringed hindwings. Body is 1 cm long with 1.5 cm wing- span. The adult moths feed on nectar and are nocturnal in nature provided with filiform antennae. Male moths survive for less time *i.e.,* upto about 20 days where as female live for 55 days or even more

Damage: The first instar caterpillars that bore and attack flower buds, flowers and bolls cause damage to the crop. They feed on seeds internally and seal the holes after entering the bolls. The young attacked bolls invariably fall down. They cause damage to ginning percentage oil content and spinning qualities. Sometimes within the cotton seeds they may undergo dipause and are transported from place-to-place. The infestation ranges from 40-90%. Most importantly the seeds from damaged bolls show low germination power hence its effect pervades longer, even in next generations of crop.

Control

i) Removal and destruction of infested shoots, fruits and fallen bolls to prevent re-infestation.

ii) Ratooning of cotton should not be practiced in infested areas.

iii) Alternative host plant growing in the surroundings of the crop should be destroyed.

iv) Seed should be fumigated by methyl bromide at 3.5 kg/100 cu.m for a day or with aluminium phosphide @ 150 g/100 cu.m for seven days before the are sown.

v) Insects should be caught by pheramone or light traps.

vi) Biological control includes, *Triphlapi pectinophorae* (egg parasite), *Microbracon lefroyi* (larval parasite) etc.

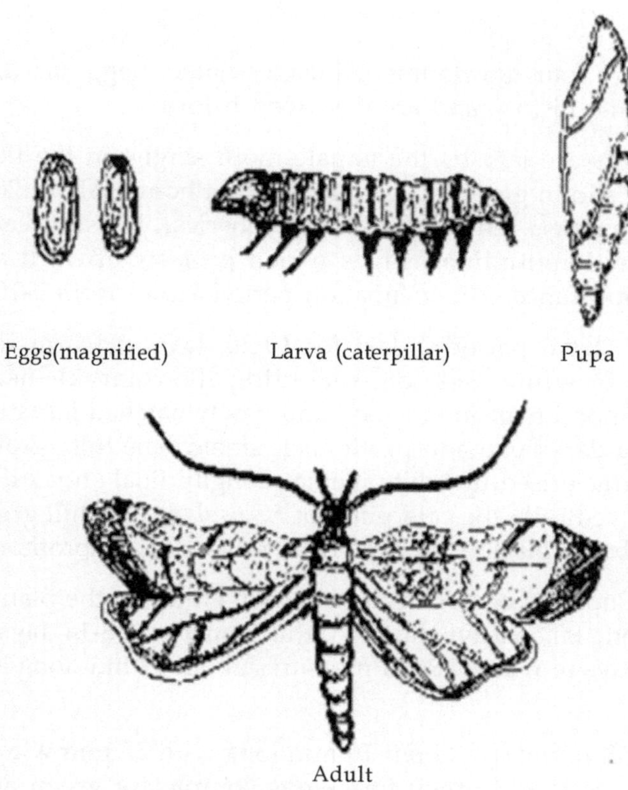

Life-cyc le of *Pectinophora gossypiella*

2. **Spotted Boll Worm**

Systematic Position

Order	:	Lepidoptera
Family	:	Noctuidae
Genus	:	Earias
Species	:	i) *insulana,* ii) *vitella*

Host plants: The main host plant of spotted bollworms is cotton, but also infests lady's finger maize and various plants of family's malvaceae and Tilaceae.

Distribution: These two species are widely distributed in countries like Spain, Syria, Palestine, Myanmar, Australia, India, Pakistan, China and some parts of Europe. In India they are mostly found in Madhya Pradesh, Maharashtra, Karnataka and Tamil Nadu

Life-cycle

There are four developmental stages namely egg, larva, pupa and adult in its life history and are described below:

Egg: Eggs are laid by the female moth singly on the flower buds, bracts, bolls and in green leaves during night hours. About 200-600 eggs are laid by a single female, which are spherical, bluish green in colour with parallel longitudinal ridges which projects upward and give it crowned appearance. The incubation period varies from 2-10 days.

Larva: Larval period is last for 10-20 days in warm regions and 50-60 days in winter session, including the 5-larval instar. Larval development occurs inside the boll and newly hatched larva is brownish white with a dark head and prothoracic shield. The fully grown larva *E. vitella* is characterised by white median longitudinal streak dorsally and pale yellow ventrally the caterpillar of *E. insulana* has dull greenish body with a number of black marks and oval, range dots on prothoracic shield.

Pupa: Pupation occurs inside a cocoon either on the plants or on the ground among fallen leaves and the adults emerge in 8-14 days in summer and 18-23 days in winter. Pupa measures about 12 mm long and 4.5 mm in breadth.

Adult: *E. Insulana* is about 10 mm long with 25 mm wing span. The body colour is bright green forewings completely green and is more prevalent in dry regions of cotton growing fields in the world.

E. vitella moths are yellow green and measure same as that of above. They have characteristic narrow light longitudinal green band in the middle of the forewings. The hind wings of the moth are white in colour and body bears tuft of hair behind the thorax.

Damage: When the crop plant is 6 week old, the larva bore into the terminal portions of the shoots, causing drooping and drying of shoot due to its feeding by boring into it. When the flower buds and bolls appear, the larva start is feeding on them. Large bolls bear holes pluggd with excreta of the caterpillars. Generally infested bolls shed and remaining ones open prematurely, resulting in poor quality thread with reduced market value. It causes 50-80%, loss of the crop during severe infestation.

Control

i) Uprooting and burning of cotton stalks soon after harvest.

ii) Destruction of alternative plants that host the pest in the vicinity of crop.

iii) The withered and withering tops of the plants that are infested should be removed and destroyed.

iv) After harvesting of the crop. The field should be ploughed to distroy the hidden larvae.

v) Complete destruction of dropped bolls, fallen cotton stalks, infested shoots and bolls.

vi) Chemical measures include spraying with carboryl 50 WP @2.5 kg/ hectare, endosulfan 35 EC @ 2.5 Liter/hectare, phosalone etc. after every 15 days.

vii) Biological controls include, *Trichogramma, chilouis* (egg parasite), *Microbracon lefroyi, M. hebets, Rhogas,* Apanteles (Larval Parasites), *Melcha mursei* (Pupal parasite) etc.

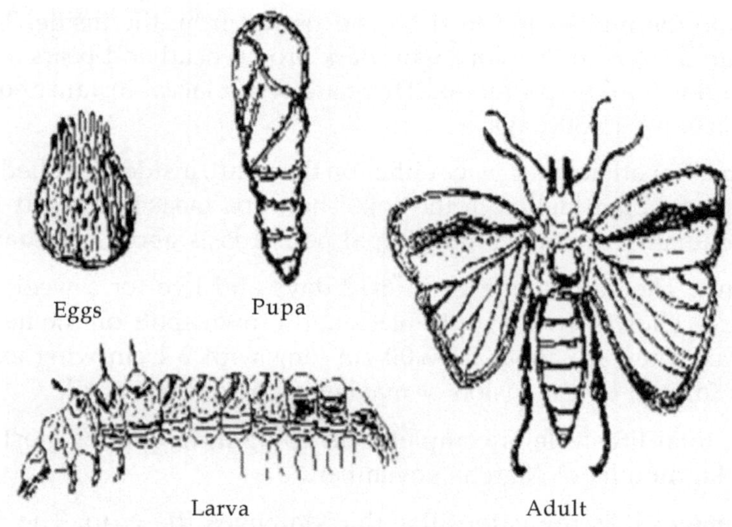

Life-cycle of *Earias vittella*

3. **Cotton Leaf Roller**

Systematic positions

Order	:	Lepidoptera
Family	:	Pyralididae
Genus	:	Sylepta (Haritalodes)
Species	:	*derogata*

Host plants: Besides cotton, it has also been found on various plants of family malvacea like *gulkhaira, okra* etc.

Distribution: The cotton leaf roller is found in various countries like India, Pakistan, Myanmar, Sri Lanka, China, Japan, Egypt, Africa and Australia etc. in India, it occurs in all the cotton growing tracts and is an important sporadic pest. It's commonly found in Madhya Pradesh, Maharashtra, Mysore, Chennai, Gujarat, Uttar Pradesh, Delhi, Punjab etc.

Life-cycle

Egg: The female moth lays 250-300 eggs singly on the underside of the leaves during the evening hours. These eggs are initially white in colour and turn to brown Just before hatching. The eggs hatch in 2-6 days.

Larva: The newly emerged larvae first feed on the lower surface of the leaves. The older larvae roll leaves from the edges inwards Uttar Pradesh to the midrib and feed on leaf tissues from the inside. The full grown larva is green in colour with dark brown head and bears a brown coloured notch in the prothorax. There are seven larval instars and larval period lasts for 15-30 days.

Pupa: Pupation takes place either on the plant, inside the rolled leaves or among the plant debris in the soil. The pupa measures about 12 mm and is reddish brown in colour. Pupal period lasts about 6-12 days.

Adult: The adults emerge in 6-12 days and live for a week. Adult moths are yellowish-white, with black and brown spots on the head and the thorax. They are about 2.5-3.00 cm long with 2-4 cm wing expense. Snout is formed by the fusion of maxiallary palps and labial.

The total life-cycle is completed in 25-55 days and its most active during the months of March–November.

Damage: It is the caterpillar that damages the crop. The freshly emerged larvae feed on cotton leaves (on their epidermis) and when they grow older they roll up. In severe infestation, the plants are completely defoilated, resulting in the decreased rate of photosynthesis which seriously effects cotton yield.

Control

i) Regular hand picking of rolled leaves minimize the attack of pest.

ii) Since, larvae are capable of hibernation under the debris and soil. Thus the field should be ploughed deeply before sowing of the crop.

iii) The seeds should be fumigated before they are sown.

iv) The damage of the pest can be reduced by sowing short duration varieties and by early termination of last irrigation to the crop.

v) Spraying the crop with chemicals like carabaryl 10% dust @ 20 kg/hectare, fenitrothion 5% dust @ 20 kg/hectare.

vi) Biological control includes *Trichogramma minutum* (Egg parasite), *Bessa remota, Elamus indicus, Microbracon* lefroyi (Larval parasites), Xanthopimpla (pupal parasite) etc.

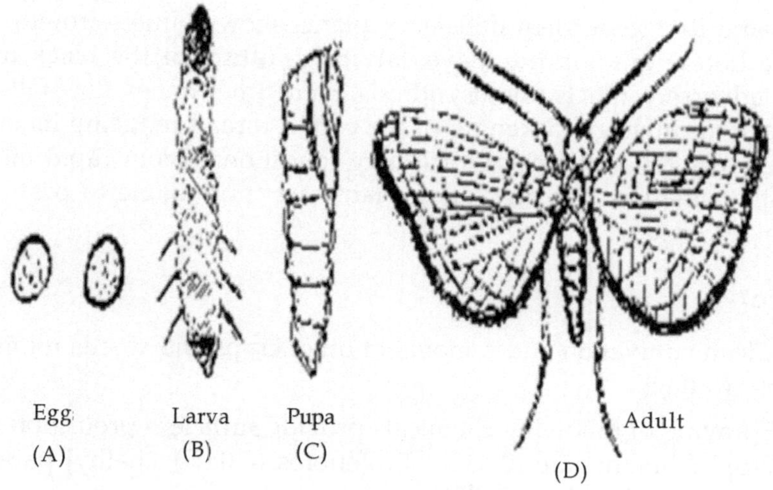

Life-cycle of *Sylepta derogate*

4. **Cotton Aphid**

Systematic Position

Order	:	Hemiptera
Family	:	Aphididae
Genus	:	Aphis
Species	:	*gossypii*

Host plants: It feed on cotton, brinjal, lady's finger, potato, chilli's pulses, ground nuts, coffee, peper, okra, guava etc.

Distribution: Its is mostly found in temprate, tropic and sub trophic zones. In India it's found in almost all cotton growing states. Since, it's highly polyphagous therefore it's also found in non-cotton growing states.

Life-Cycle

The life-cycle of *Aphis gossypii* is very complicated. The adults exist in both winged (alatae) and wingless (Apterae) forms. The adults are small, greenish brown and soft bodied found in colonies on the tender parts of the plants and under surface of the leaves. They are viviparous and capable of parthenogenesis to increase their number. About 8-20 nymphs are produced in 10-15 days. There are four nymphal instars and nymphal period lasts for 7-10 days. Many generations are completed in a year, with one generation taking a time period of 8-12 days.

Damage: The damage is caused both by nymphs and adults; they suck plant sap from the lower surface of leaves causing curling of the leaves and due to nutrient deficiency plants show stunted growth. They excrete honey dew on the leaves inviting culture of the black mould which adversely affects photosynthesis. Sometimes honey dew falls onto open cotton causing blackening of the cotton thread reducing its quality and growers get a little price for it. Dry conditions favour rapid increase in pest population and the young plants are more liable to pest attack than older ones.

Control

i) Clean cultivation and removal of other crops and weeds minimizes pest attack.

ii) Spraying of following chemicals provide sufficient protection to the crop. Dimethonate (0.03%), Profenofos (0.05%), methyl parathion (0.025%), phosalone (0.05%) etc.

iii) Biological control includes, predators like *coccinella septem punctata, syrphus confractor, Triphiles tantilus,* chrysopate.

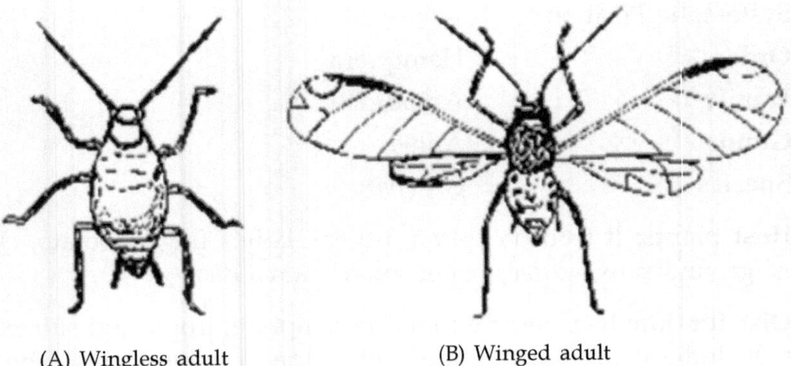

(A) Wingless adult (B) Winged adult

Cotton aphid *Aphis gossypii*

19

Sericulture, Lac Culture and Apiculture

Silk is a way of life in India. Over thousands of years, it has become an inseparable part of Indian culture and tradition. No ritual is complete without silk being used as a wear in some form or the other. Sericulture and Silk Textiles Industry is one of the major sub-sectors comprising the textiles sector. Sericulture is an agro-based cottage industry. Sericulture refers to the mass-scale rearing of silk producing organisms in order to obtain silk. Sericulture is an agro-based labour intensive industry. The major activities involved in a sericulture industry are:

1. Cultivation of silkworm food plants
2. Rearing of silkworms for the production of raw silk
3. Reeling the cocoons for unwinding the silk filament and
4. Other post-cocoon processes such as twisting, dyeing, weaving, printing and finishing.

Sericulture is one of the most labour intensive sectors, combining activities of both agriculture (sericulture) and industry. India is ranked as the second major raw silk producer in the world. It is this position along with its immense employment potential that makes sericulture and silk, indispensable in the Indian textile map. Silk is a high value but low volume product accounting for only 0.2 % of world's total textile production. It churns out value added products of economic importance.

India holds the monopoly on producing the Muga silk. It is the only one cash crop in agriculture sector that gives returns within 30 days. Sericulture emerged as an important economic activity, becoming increasingly popular in several parts of the country, because of its short gestation period, quick recycling of resources. It suits very well to all types of farmers and exceptionally for marginal and small land holders as it offers rich opportunities for enhancement of income and creates own family employment round the year.

The Government of India (GOI) has concurrent responsibility for the development of the silk industry in the country. At the Central level, Ministry of Textiles is the nodal organisation with 'Sericulture and Silk Textiles Industry' as one of the principal functional areas.

There are several centrally sponsored schemes for promotion and development of sericulture sector, through which Government of India has been undertaking different activities like creation of sericulture related infrastructure; development of nurseries and farms; expanding plantation areas; etc.

Family	Species	Host plant	Country
Bombycidae	*Bombyx mori*	Mulberry	South-East Asia
	Bombyx spp.	Mulberry	India
Lasiocampidae	*Gonometa postica*	*Acacia* spp.	Africa
	Trabala vishnou	Polyphagous	South-East Asia
Saturniidae	*Actias selene*	Many Hosts	India to China
	Antheraea assamensis	Polyphagous	India
	Antheraea mylitta	*Syzygium* spp.	India

SILK

Silk is a valuable product and due to its elegance and colour, silk fibre is rightly called "Queen of Textiles". It's the secretion from the salivary glands commonly called silk glands of larval silkworms. These glands are found on both sides of the alimentary canal of silk worm larvae and this secretion hardens into fine threads (silk). The cocoons enclosing pupae form the raw material from which thread is extracted. Most of the caterpillars of order lepidoptera form cocoons bearing extractable silk threads but those of only two families i.e., bombycidae and saturnidae produce silk of practical use

Types of Silkworms: Some of the important silkworm species are described below:

i) **Muga Silkworm *(Antheraea assamensis)*:** It is a wild and semi domesticated species in Assam, West Bengal, Bihar and Odisha. The caterpillars feed on the leaves of Cinnamon and matchilis plants. Cocoons are amber or white and silk obtained after reeling is golden yellow in colour.

ii) **Tussore or Tassar or Tropical tasar silkworm *(Antheraea mylitta, A. paphia)*:** It's found in China, India, and Sri Lanka etc. In India it's found in the forests of Bengal, Assam andUttar Pradesh it remains still undomesticated and its larvae feed on leaves of fig, oak, ber, sal etc. Cocoons are spun by the final instar larvae and are oval and hen's egg size, brown red or yellow with hard case hung from the terminal braches by a stiff attachment. These have to be collected from the forests. The silk is reelable and the pupae have to be killed before emergence of the adult that cuts the silk thread into pieces.

iii) **Eri silkworm *(Philisamia richi)*:** It's found in South East Asian countries including India. It feeds on castor leaves and produces a rough and strong silk locally known as "Arandi Silk". This silkworm is mainly found in the forests of Assam, Bihar, Bengal, Uttar Pradesh, Odisha and Madras. Cocoons are white in colour and unreelable therefore produce inferior quality silk. Despite being dull in colour yet it has long durability.

iv) **Mulberry Silkworm. *(Bombyx mori)*:** It is the most important and commonest of all which produces most of the raw silk we use Bombyx mori is a native of China and almost domesticated worldly. It feeds on mulberry (Morus Alba) leaves for so many centuries that it no longer exists in wild state. By careful selection and hybridization many races have been developed to meet the various needs of climate, quality and quantity of the silk obtained. The silk obtained from this silkworm is white or yellow in colour. There are several races of the mulberry silkworm which may be grouped into two categories viz. univoltine and multivoltine. The univoltine has an annual life-cycle and produces superior quality of silk while multivoltine race passes through many generations in a year and silk produced is of inferior quality.

Biology of Bombyx mori

Systematic Position

Phylum	:	Arthropoda
Class	:	Insecta
Sub-class	:	Pterygota
Division	:	Endopterygota
Order	:	Lepidoptera
Family	:	Bombycidae
Genus	:	Bombyx
Species	:	*mori*

Habit and Habitat: Silkworm is purely domesticated from a very long past and can no longer survive *n*ature without human help. The larvae have lost the capacity to search food and will die of starvation unless food is placed right near them. Commercially they are reared on mulberry leaves.

Morphology: Adult silk moth is a medium sized insect about 26 mm long with 40-50 mm wing extension. The male is smaller than female and body is robust, creamy-white or yellow in color and divided into 3-parts: Head is small and bears a pair of black compound eyes and bi pectin-ate antennae and degenerate siphoning mouth parts. Thorax bears three pairs of small legs and two pairs of large but feeble wings marked by several faint or brown lines. Entire body is covered with hairs. Abdomen of female is larger and much distended due to great number of eggs present in it. The moths copulate immediately after emergence by keeping their mouth to opposite directions.

Life History: Silkworm is a typical dioeciously insect undergoing internal fertilization followed after copulation. The various developmental stages include :

1. **Eggs:** Soon after the fertilization each female lays about 300-500 eggs in the night in clusters on the undersurface of the mulberry leaves. Female covers the eggs by a gelatinous secretion which holds the (eggs) with leaves. The eggs are small, smooth, spherical, pale white and seed like in appearance. After laying eggs, female does not take food and die within 4-5 days of starvation. The eggs laid are either non diapauses type (tropical countries) or diapauses type (temperate countries). During summer they hatch in 10-12 days and in 30 days during winter.

2. **Larva**: The larva which hatches from egg in about 10 days is known as the caterpillar. The caterpillar on hatching refereed as 1st instars larva is a tiny creature about 4 mm long, white to dark in color, moves about in a characteristic looping manner. It has a rough and wrinkled body which is made of 12 segments. There are 3 pairs of thoracic legs and 5 pairs of abdominal legs which are provided on the 3, 4, 5, 6 and 10th abdominal segments. Dorsal spine is found on the dorsal surface of eight abdominal segment its head bears manbulate mouth parts with which it at once starts feeding on mulberry leaves and grows very quickly young larvae are kept at 25-27°C temperature. After 4 or 5 days, it stops feeding and becomes inactive. Then ecdysis or moulting takesplace, older skin bursts and a new skin and slightly bigger in size than the older caterpillars. This larva resumes eating voraciously and growing untill the skin is

again cast off after a week. The larva repeats this process four times so that there are 5 instars innolved in the development.

The fully grown 5th instar larva is creamy white in colour; 21-25 days older about 75 mm long and weights about 5 Gms. The description of which is given in the beginning under sub-heading larva. Its small head bears bitting and chewing mouth parts and a series of respiratory spiracles or ostia are provided on both Silk gland complex and silk gland regions ateral sides.

Tassar Silkmoth

Eri Silkmoth

Muga Silkmoth

A pair of salivary glands

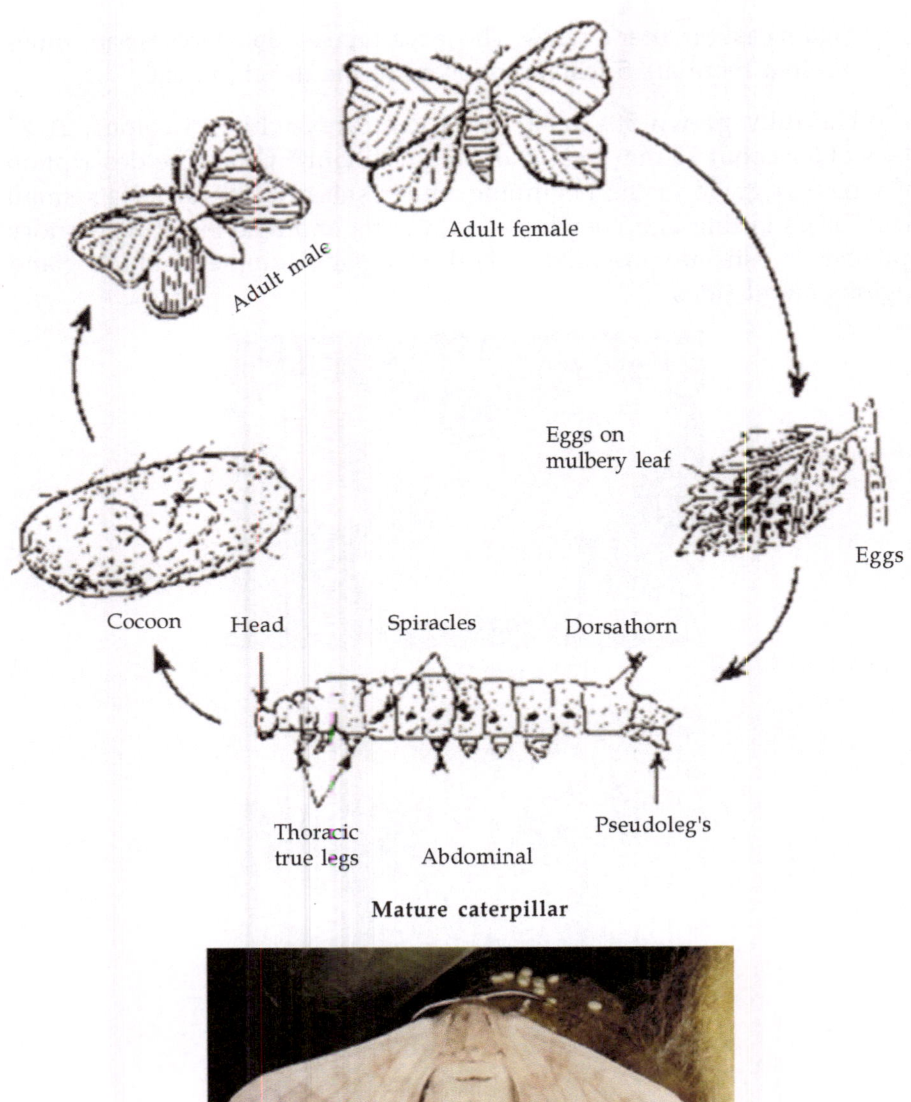

Mature caterpillar

Bombyx mori (Silk moth)

Stages of life history of Bombyx mori develop in the lateral side of its body. In *Bombyx mori* silk glands are highly developed and are enormously covering hind region of the gut. These are about four times longer than the actual body length. These glands are homologues to the salivary glands of other insects. Interiorly each gland opens into a long salivary duct. Both salivary ducts unite apically and open at the apex of the median cylindrical structure known as the spinneret. Each such silk gland is differentiated into three parts. Anterior part consisting 250 cells and 3 cm long, middle part is 7 cm long consisting 300 secretary cells and the posterior part is 15 cm long consisting 500 secretor cells. Massive secretion occurs during the last four days of the fifth instars larva. The silk protein is composed of 60-70% fibroin and 20-25% sericin, which are secreted by posterior and middle part of the silk gland respectively.

3. **Pupa or Chrysalis**: Mature caterpillar stops feeding and returns to a corner among the leaves. Its salivary glands now secrete a sticky fluid through a narrow pore called spinneret situated on hypo pharynx. The sticky substance turns into a fine, long and solid thread of silk as it comes in contact of air. It becomes wrapped around the body of caterpillar forming cocoon. The cocoon is 38 mm in length and 19 mm in breadth, oval, yellow or white in color. A single caterpillar is said to produce nearly 1,000-1,500 meters of silk thread in this manner. Pupation occurs inside the cocoon and last for 10-15 days. To obtain good quality of silk the pupae or chrysalis are not allowed to get transformed into adults and are killed inside the cocoon—stiffing. Active metamorphosis takes place during pupation. Some of the important changes include disappearance of unsegmented abdominal prolegs, development of two pair of wings on thorax part. It secretes an alkaline fluid to moisten one end of the cocoon and forcibly makes its way out of it.

Method of Rearing Silk-worms

The major steps and requirements of sericulture industry are briefly as follows:

1. **Cultivation of Mulberry Plants:** Cultivation of mulberry plants to obtain leaves is called moriculture. The mulberry silkworm mainly feeds on green leaves of mulberry plants belonging to the genus Morus of family Moraceae. Among 20 species of mulberry, In India

There are three leaf harvesting methods

a) Individual leaf picking. Leaves are singly handpicked, tender leaves are provided to younger larvae and mature leaves to older larvae.

b) Branch cutting. The entire branch is cut and offered to 3rd instars larvae. The leaves are not detached but they eat up them from the branches it.

c) **Top or whole shoot:** Harvesting the fourth and fifth instars larvae is given the top portion of the shoot bearing branches and leaves. The tops of the shoots are clipped and given to them.

2. **Rearing of silkworms**: The silkworms are reared in places which should avoid dampness stagnation of air, Exposure to bright sunlight and strong winds. The temperature should lie between 25-30°C and humidity should never drop below 70% RH and rise 80% RH proper ventilation should be ensured for efficient productivity.

Accessories

i) *Rearing stands:* These are wood or bamboo stands of frames on which rearing trays are placed.

ii) *Rearing trays:* Rearing trays are made up of split bamboo used for rearing caterpillars Since, young larvae i.e., 1st and 2nd instars larvae are very delicate and susceptible paraffin paper is used to cover bottom and top of the rearing trays which are generally circular but sometimes box type wooden trays are also used. Paraffin paper are used to maintain humidity and prevent withering leaves.

iii) *Ant wells:* These are circular or rectangular water containing bowls in which legs of rearing stand are kept to prevent attack of ants.

iv) *Chop sticks:* These are tapering bamboo rods required for picking up younger larvae to avoid any injury.

v) *Feathers:* These are usually white feathers used for brushing together freshly hatched larvae to prevent injuries.

vi) *Baskets:* Baskets made of bamboo are needed to fetch mulberry leaves.

vii) *Chopping knives:* These are required to cut the mulberry leaves into fine pieces.

viii) *Chopping boards:* Chopping board on which leaves are cut is made up of soft wood and placed on mat.

Procedure of sericulture: The seed cocoons are reared under optimum conditions in the drainage and for obtaining seeds they are spread as a single layer on the trays arranged in racks. The place is well ventilated and a constant temperature of 23-25°C and humidity between 75-80% RH is maintained. Once the adults emerge after 10-12 days of cocoon formation, the two sexes are separated and allowed to mate in the same way as discussed earlier. Then eggs are to be collected for which there are several methods such as cellular bag method, cellular card method, flat card method etc. In cellular card method, a craft paper is divided into 20 squares in 4 rows each having 5 compartments. The mated females are kept on the card squares under moth funnel. Once the eggs are laid, the female moths are removed. The eggs are soaked for 30 to 60 minutes to separate the eggs from the paper which are again soaked for 10 minutes in 0.5% of bleaching powder solution to remove the glue. Then they are transferred to a salt solution of 1.06-1.20 sp. gravity to remove floating unfertilized eggs. Then the fertilized eggs so obtained are washed in 2% formalin to disinfect them. Then they are dried and packed in boxes for marketing.

Stiffing: For obtaining the commercial silk, the cocoons are treated with hot water or placed in a hot oven so that pupae are killed inside the cocoons. Killing of pupae is economically important to prevent the emergence of moth that cuts the cocoon and makes that unreliable.

Reeling: Removal of silk thread from cocoon is refereed as reeling. The cocoons are first heated in boiling water to soften them and dissolution of sericin or gum occurs. It's followed by brushing the outer surface of cocoons so that free end of the silk filament is obtained. In filature method of reeling the free ends of the silk filament of 5-10 cocoons are picked together, fixed on the reeling appliance and all are twisted into a single thick thread and wound on a reel. More importantly in this method cocoons are subjected to steam bath and not the boiling water. The thread is continuous and extremely long, the silk so obtained is called reeled silk. The raw silk is first reeled on small reels that allows the proper drying of.

Fiber: Then its reeled on large reels refereed as reeling. Only about 50% of the silk in each cocoon is releable, remainder is used as silk waste and formed into spun silk. Like many other insects, silk worms too are susceptible to pathogens causing them ill and eventually leading to their death. Some of the important among them are:

Pebrine: It's the serious disease of silkworm caused by protozoan Nosema bombycis. The disease is transmitted through contaminated food and contact. It also spreads through the eggs of the diseased moths. The spores once ingested invade the gut tissue and are continuously discharged through the fecal matter. The infested moths show low fecundity, produce pepper like spots on body, wrinkled skin and sluggish behavior. Pebrine is of two types, one protozoan type already discussed and other is called viral pebrine caused by the virus namely Bo reline bombycids, the larvae suffering from this disease get killed in 8-10 days of infection.

Flacherie: It's a bacterial disease caused by *Bacillus thuringiensis* sotto. The diseased larvae vomiting green fluid through their mouth and are generally thin in body musculature. Due to high dysenteric condition larvae appears flabby, feeble, weak, withered and loosely hanging. The body putrefies and becomes black or green. The caterpillars die and their bodies give an offensive smell. This disease is generally caused by indigestion, therefore over feeding should be avoided. Regular feeding of the larva and maintaining good hygienic conditions prevent the spread of disease.

Muscardine: It's a fungal disease caused by fungus, Beauveria bassiana and commonly found during rainly season. The fungal spores adhering to the larval body germinate under suitable conditions and penetrate the body. The suffered caterpillars loose appetite, become soft bodied turn stiff. White muscardine is due to Beauveria bassiana green due to spicaria parsinna and yellow muscardine via Iscaria farinosa.

Glasseria: Its caused by barreling virus, causing swelling of body segments that ultimately results in bursting of skin, care should be taken that the larvae should not be fed on mature leaves first and tender leaves after wards. Use of resistant strains, chemical disinfectants like bleaching powder, formalin, etc. should be periodically used to check the infestation.

Sericulture in India: In India, Lefroy (1904-05) was first to investigate on silkworm and sericulture at Pusa Institute, New Delhi. But the planned development of this industry was taken UP only after the establishment of the silk board in 1949. Today sericulture provides longer oppurtunities for employment of youth. India is the only country known for producing all the four commercial varieties of natural silk. The main research and training centers are:

- Central Sericulture Research Stationin Behrampur, West Bengal.
- Central Sericultural Research and Training Institute, in Mysore.
- Central Tasar Research Station Ranchi, Jharkhand etc.

These centers are working under Central Silk Board (Ministry of Textiles, Government of India) and provide assistance to silk growers, throughout the country mostly in Karnataka, Punjab, J&K., Odisha, Tamil Nadu, Mysore, Jharkhand, etc.

Morphological differences Butterflies and Moths

S.No.	Butterflies	Moths
1. Antennae	**Butterflies and Skippers**: Have a thickened club or hook on the tip of the antenna, never 'feathery'	**Moths**: Have simple thread-like or 'feathery' antenna without a club
2. Colour	**Butterflies and Skippers**: Brighter colours	**Moths**: Duller colours
3. Wings	**Butterflies and Skippers**: Wings are not linked – no frenulum	**Moths**: Wings are linked together with a bristle-like structure called a frenulum
4. Resting posture	**Butterflies and Skippers**: Hold wings together above body when resting	**Moths**: Hold wings flat when resting
5. Forelegs	**Butterflies and Skippers**: Forelegs reduced, missing terminal (end) segments	**Moths**: Forelegs fully developed
6. Pupae	**Butterflies and Skippers**: Pupae (chrysalids) not in cocoon	**Moths**: Pupae spin a cocoon
7. Activity	**Butterflies and Skippers**: Fly during the day.	**Moths**: Fly at night:

Butterfly

Moth

LAC-CULTURE

Distribution and Host Plant *Laccifer lacca (Tachardia lacca)*

Lac insect *Kerria lacca* (Khair), *Acacia arabica* (Babul), *Ficus religiosa* (Pipal) is distributed in India, Sri Lanka, Java, Malaysia, China and Thailand. The important lac producing countries are India 62% of the total world out put. At present it is used in large number ofindustries, furniture and floor polishes, paints etc.

Phylam	:	Arthropoda
Class	:	Insecta
Order	:	Hemiptera
Super-family	:	Coccidae
Family	:	Lacciferidae
Genus	:	Laccifer
Species	:	*lacca*

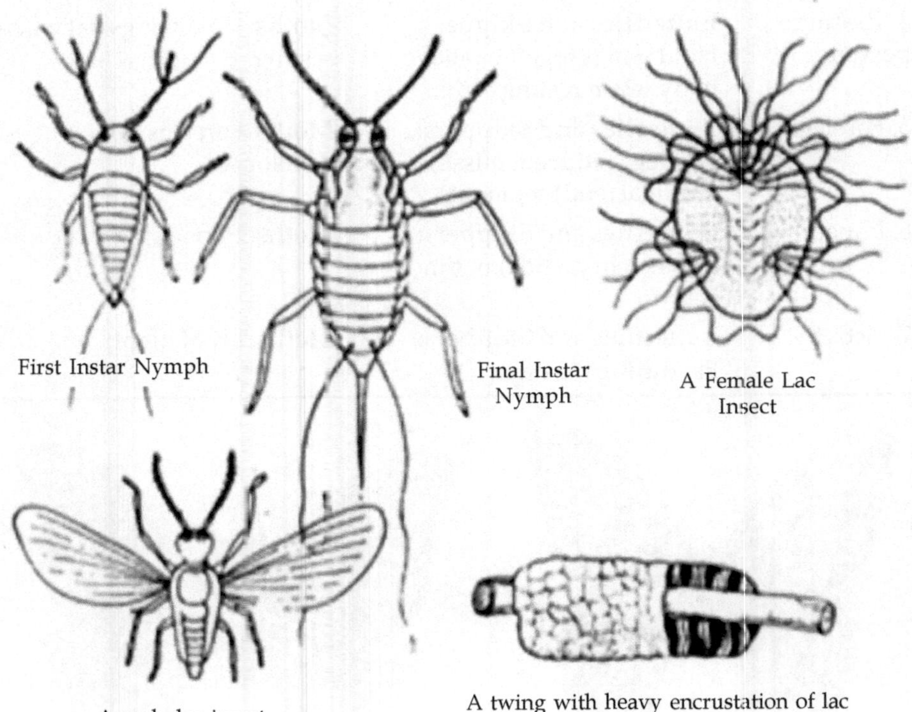

Tachardia lacca

Male Lac Insect: The head is a large prominent mouth parts but vestigeal. Ocelliare two pair present, abdominal seven segmented antenna which has hair and segmented. Tarsi are 3 segmented, abdominal narrow and last segment bears pointed penis. They are pinkish red colour and are two type winged and wingless insect. The wing male translucent membrane forewing. They are mostly found during dry season. They survive only 3-4 days and die after copulation lac insect.

Female Lac Insect: The female ventral surface of the lac insect body is flat and dorsal surface is convex. The antennae are vestigeal small and 3-3 segment. The mouth parts are piercing and sucking type, rostrum 2 segmented. The female remains attached at a point and sucks the juice from the host spiracle are open and present on mesothorax. The eyes and legs and wing deprive the female. Abdominal segment is two segmented and anus is fringed. The development stages found in its life cycle include eggs, nymph and adult. The lac female eggs are laid by the female within the lac-cells. It lays about 200-400 eggs which may be fertilized or unfertilized eggs. Both types of eggs give brith to male and female nymphs. The eggs from which both male and females hatch out in equal number. From which both males but less than female are born after sometime. The female generally lays well developed eggs which hatch within 2 hours that is why some scientists call them viviparous female. The newly hatching nymph is 0.4 × 18 mm long and pinkish in colour. The lac insect antennae possess three pairs of legs, two compound eye and six anal setae.The lac nymphs emerge in large numbers and crawl over the surface of twigs and branches of plants in search of suitable piece for suck. They mostly settle on shoots gregariously after two days of setteling they start secreting resin from glands distributed under the cuticle all over the body except near the mouth parts and the anus. Thus, the nymphs gets incased in a cell of its secretion which increases in size with the growth of the insect. They moult thrice before reaching maturity and after 1st moulting the nymph loses its eyes, legs and Antennae. The anal setae become 4 instead of 2. The male nymph develops after second moults. The male and female continue to develop from the male cells both wing and wingless male emerge after two months. These males fertilize the females in their cells and after fertilization female develops very fast and secretes the lac at a faster rate. Thus the female lac insect are the chief source of lac secretion.

Lac culture of insect

The lac culture company is well developed in India and at present it accounts to about 70-73% of total word production the lac produced by different state of the country is given below. Bihar 41% M.P. 26.5% and W.B. 20.6% U.P. little production of India.The former is raised mainly on the Kusum tree in the host plant of lac insect. The two methods of lac cultivation, one method of local and second scientific methods. 1. The local methods of lac cultivation in which lac is collected from the trees growing in jungles. This method of lac cultivation has generally failed.

2. **Scientific methods:** This method was developed by Indian lac culture R.I. Ranchi and is known as coupe system. In this system as

against the local practice of keeping all the under tree continuous cropping a certain number of tree is a coupe or compartment are inoculated fully and when lac matures, is fully reaped and another set of trees from another coupe is taken up for inculation of lac culture.

i) *Selection of Suitable Site of Insect Lac Culture:* The site should be selected on the basis of host plants, where environmental conditions suitable for the host tree Kusum plant should be preferred. The excessive heat and cold should be avoided and rainfall may be about 20 mm for host plant tree.

ii) *Inoculation of Broad Lac Culture:* The broad lac sticks should be cut into convenient length of 10-14 cm and tied in succulent shoots of the host tree. When the broad lac is removed from the tree its is known as Phunki lac. There are following types of lac.(i) Ari lac, (ii) stick lac, (iii) seed lac, (iv) dust lac, (v) shel lac(iii) Scrapping of Broad Lac Insect.

Commercial Products of Lac Insects

1. Aleuritic acid is a very valuable compound used as a starting raw material for the preparation of the chemical used in pharmaceuticals drugs etc.
2. Insulating varnishes kinds of shellac based air drying and baking type insulating varnishes having high termal resistance have been developed.
3. Lac dyes and lac waxes are important products used in other industries.
4. Surface coating melfolac, polish primer paints etc.
5. Binder adhesive in the book and the gasket-shellac compound formulation has been developed for automobiles. The slow release composition for control of cockroaches has also been developed in the lac product.

Cultivation of Lac

Cultivation of lac involves proper care of host plants, regular pruning of host plant, infection or inoculation, crop-reaping, control of insect pests, and forecast of swarming, collection and processing of lac. The first and perhaps the most important prerequisite for cultivation of lac is the proper care of the host plant. It is the host plants on which lac insects depend for their food, shelter and for completion of their life cycle. There are two ways for the cultivation of host plants. One is that plants should be allowed to grow in their natural way and the function of lac-culturist is only to protect and care for the proper growth of plants.

Another way is that a particular piece of land is taken for the purpose and systematic plantation of host plant is made there. Regular watch is necessary in this case by providing artificial manures, irrigation facilities, ploughing and protecting the plants from cattle and human beings for which the land should be fenced. The larvae of lac insects are inoculated on host plants only after the host plants have reached a proper height. The lac larvae feed on the cell sap by inserting their proboscis in the tender twigs. The proboscis can only be inserted in the tender young off-shoots. For this before inoculation, prunning of lac host plants is necessary. The branches less than an inch in diametre are selected for pruning. Branches half inch of less in diametre should be cut from the very base of their origin. But the branches more than half inch diametre should be cut at a distance of 1 ½ inch from the base.

Lac Enemies and Their Control

A lac enemy imposes a challenge to the lac culturist, as they not only decreases the population of lac insects, but also retard the production and quality of lac. Damage caused to lac insects may be grouped under two heads, (a) damage caused by insects (b) damage caused by animals other than insects. Insect enemies of lac crop may be predators and parasites. The common parasites of lac insect are known as "Chalcid." They are small, winged insects which lay their eggs inside the lac coat either on the body of the lac insect or inside the body of the lac insect. The larva which hatches from these eggs feed upon the lac insects, thereby causing mortality of their host. Damage done by this parasite constitute about 5-10% of the total destruction of the lac crop. Damage done by the predators is of greater intensity (35% of the total destruction). The major predators of lac insects are *Eublemma amabilis* (the white moth) and *Holococera pulverea* (the blackish grey moth). They not only feed on lac insects but also destroy the lac produced by term. Squirrels, monkey, rat, bat, birds (wood peckers), man etc., are the enemies other than insects which destruct the lac crop in different ways. Damage is also done by climatic factors such as excess heat, excess cold, heavy rain, and storm and partly by the faulty cultivation methods.

Use of Lac

Lac has been used for the welfare of human beings from the great olden days. No doubt the development of many synthetic products have made its importance to a little lesser degree, but still it can be included in the list of necessary articles. Lac is used in making toys, bracelets, sealing wax, gramophone records etc. It is also used in making grinding stones, for filling ornaments, for manufacturing of varnishes and paints, for silvering the back of mirror, for encasing cable wires etc., Waste materials produced during the process of stick lac is used for dying purpose. Nail polish is a good example of the by-product of lac.

Importance of Lac Culture

1. Lac cultivation is a good source of livelihood for poor tribals inhabiting sub-hilly tracts with meagre investment.
2. It is a good scope for marginal and degraded lands and has no competition for land opera-tion with agriculture/horticulture crops.
3. Highly remunerative cultivation. For exam-ple, one hectare of Ber plantation with Kusmi lac cultivation can produce net return of Rs. 3-5 lakh/year.
4. It is like an insurance crop especially during drought year as the crop is very good during such adverse climate.
5. A good number of host trees like Ber, Palas, etc. occurring naturally in forests and sub-forests in Chattisgarh are available for commercial exploitation.
6. It is a good source of employment generation. One hectare of Kusmi lac cultivation on Ber generates 620 man-days of employment/year.
7. Lac is the most superior natural resin. Besides resins, lac insects also yield lac dye and lac wax, all of which find extensive uses in Ayurveda and Siddha systems of medicines.
8. Lac culture can help in ecosystem develop-ment. Cultivation of lac not only provides livelihood to millions of lac growers but also helps in conserving vast stretches of forests, lac insects and associated biota. Most of the lac host plants grow in forest areas and farmers resist felling of these trees, thereby protecting them for lac cultivation. Thus, lac culture plays a vital role in the protection of our bio-resources. Cultivation of lac host plants for timber and fuel yields revenue in cycles of long years, whereas cultivation of lac on these trees gives a return almost every year.
9. It has high export potential about 75%. In the world, India is the leader in lac production and export.
10. Lac cultivation involves significant women participation.

APICULTURE

Different species of Honey bees

Species of Honey bees: Honey bees are the most familiar members of the order Hymenoptera. These are characterized by having two pairs of membraneous wings, chewing and lapping type of mouth parts and a sting in the females (workers). There are eight-species of honey bees in Asia. They include:

Apis dorsata; Apis florea; Apis cerana; Apis laboriosa: Apis nuluensis; Apis nigrocincta; Apis koschernikovi; Apis andreniformis. Apis mellifera is the lone species found in Europe, but it has obtained commercial status across the word. India is the unique country where four species of honey bees have been found foraging under natural conditions. Among them two are wild and can't be domesticated *(A. dorsata* and *A. florea*) and two are docile in nature and easy for subjugation (*A. cerana* and *A. mellifera*). Thus, for starting an apiary, it's important to select a good variety (species) of bees for domestication. Some of the obligatory characters that these insects (bees) should posses are:

I. They should be docile in nature and necessarily colonial in habitat: (ii) They should be able to protect themselves from natural and other enemies. (iii) They should be capable of producing high yield. (iv) They should adapt themselves anywhere. Four species of Apis (honey bees) commonly found in India are :

i) *Apis dorsata:* It's commonly known as Rock bee or Gaint rock bee. In size it's largest (20 mm in length) among all and its yield is maximum with 35 kg/comb. This bee is found in all parts of India and it constructs large combs of 6.9 × 15 metres. On the tall trees, buildings and caves. Tribal populations in several parts of the country harvest honey and bees wax from rock bee colonies. Despite being most efficient honey gatherer it could not be domesticated due to ferocious and irritable nature. These bees are migratory in nature. In winter, they migrate to plains and come back to hills during summer. Its an excellent pollinating agent of various wild and domestic plants

ii) *Apis florea:* It's also called the little bee, its smaller in size (8 mm) and its honey yielding capacity too is low about 250 gm/comb. It constructs small combs across 10.24 cm on branches of trees, bushes or under walls of buildings. Since, it doesnot like darkness and forms combs in open fields, that is why could not be easily domesticated. Moreover from economic point of view they are futile to be domesticated due to their low yield. Its an inhabitant of plains and can't tolerate cold, therefore not found in hills. Though its yield is low but the honey produced is sweetest one. Moreover due to its passive nature and rarely stinging property the removal of honey from combs is much easier.

iii) *Apis cerana:* It's also called the Indian bee or Asian bee or Eastern honey bee. This bee is found everywhere in India and its quite mild in temprament and can be easily domesticated. Unlike, Apis dorsata, it constructs many parallel combs in the cavities and hollows of old trees, caves and similar other sites. These bees are about 15 mm long and their yield is 3-4 kg/comb. Generally they donot migrate to leave their old combs. It responds easily for domestication and its keeping is now days practiced all over the country.

iv) *(Apis mellifera):* Its the common 'Temperate European hive bee' generally found in Europe and Italy. In India it was introduced by professor A.S. Atwal in 1962. Currently its domestication has spread throughout the country. By nature, its similar to Apis cerana but the queen is more profilic, swarms less and have good honey gathering qualities. The average honey yield of A. millifera is 20-25 kg/colony but can increase to 60 kg per colony underefficient survillience. Due to its high yield and docile nature, it's slowly replacing A. cerana from Indian apiaries. In states like Jammu and Kashmir, Punjab and Haryana. It thrives welland is an important species domesticated currently.

Bee Keeping Equipment and Methods

A. Old or Indigenous Methods

These methods include primitive and crude methods of bee keeping and can be studied under following headings:

I. *Fixed type:* As the name it indicates that this method was immovable one. This was very crude and unplanned one in which bees makes their natural fixed combs on walls or the branches of trees. Thus providing a receptacle in the wall of the house with an entrance and observation holes.

II. *Artificial movable type:* These were prepared of hollow wooden logs, empty boxes, earthen pots etc. For collecting honey by this method, the bees were often disturbed by smoke or were even killed by fire etc. Some of the important drawbacks of indigenous methods are:

i) It was a matter of chance, whether the bees get inhabited or not in the boxes.

ii) There was no direct control over the activities of bees.

iii) Honey extracted was unhygenic containing larvae, pupa, pollens etc.

iv) Since, honey extraction requires mass killing of bees hence it destroys the future yield.

v) These hives were susceptible to attack by enemies and climatic factors.

vi) Chances of improving yield and race were not possible.

B. Modern Methods

This method is also called frame hive method or longstroth's method named after its inventor (Langstroth, 1851). This method has converted apiculture into a cottage industry that provides employment to lakhs of village and urban people. In this method bees are rared in movable artificial hives brief description of the various tools used in it are given below: A standard langstroths hive accomodates 7-11 frames and is composed of:

a) *Stand:* Its a four legged structure that provides support to the whole hive. The legs are 15-21 cm in height.

b) *Floor board:* It is a tray of 55 × 24 cm dimension that rests on the stand. A floor board has the front side of 12 × 20 cm and a space at middle as the entrance gate for bees.

c) *Brood chamber:* The brood chamber is a rectangular box of 45 × 40 cm dimensions without top and bottom and rests on the bottom board. The number of brood boxes could be increased to 2 when the population increases. It internally consists of movable wooden frames arranged in a parallel fashion.

d) *Hive frames:* These are wooden frames provided with a top bar, a battom bar and two side bars. Both ends of the top bar are extended to rest on the rabbet, scooped on the long sides of the brood chamber. The under surface of the top bar is grooved to hold the comb foundation.

e) *Queen excluder:* The distance of 0.96 cm is maintained between the adjacent frames called bee space. The side bars bear 4 orifices for wiring the frames. A sheet of bees wax bearing hexagonal prints is fixed in the space of a frame and serves as comb foundation. In future the bees construct new cells on it. There is a zinc or wire frame with 2.3-3.5 mm perforations in between brood and super chamber. Since the size of worker bees lie in this range so they are free to move from one chamber to another but queen fails to do so because her thorax is 4.3 mm in thickness, its called queen excluder. To obtain pure honey, its necessary to seprate the brood chamber from the super chamber where honey is stored.

f) *Supers:* They are also called honey chambers; they are of the same size of the brood chamber and contain internally the frames of similar size and a like arrangement. They are occupied by workers and remain inaccessible to the queen. A hive could have 2-3 supers depending on its population. These are store house of wax and honey.

g) *Covers:* The top of the hive is closed by covers. There are two covers, the inner cover 50 × 40 cm is a flat board and outer hood like (roof) covers the hive. The inner cover isprovided with the exit for bees and ventilation is provided via roof.

Apiary Equipments

i) *Bee veil:* Its used as shield by the keeper to protect his face. It is made up of frames covered on the four sides with small mesh or mosquito net, top and bottom with cloth.

ii) *Bee gloves:* These are rubber gloves and during handling the bees, they provide protection from stinging.

iii) *Honey scraper:* Its a flat iron plate with one end broad and sharpened other end bearing handle generally wooden type. It is used for scraping bee glue and combs.

iv) *Smoker:* Its a tin container from which smoke is generated to drive the bees out of the supers prior to honey extraction. Moreover sometimes bees get furious, releasing smoke helps bee-keeper to calm down them.

v) *Wire embedder:* Its a tool used to embed the comb foundation (wax sheet with haxagonal prints) on the frames.

vi) *Bee brush:* It's used to brush bees from a honey comb before extraction.

vii) *Honey extractor:* It's a large drum of tin containing revolving rods bearing pockets of netted cloth. Extraction of honey is done on the principle of centrifugat forces. Comb foundations full of honey are kept in pockets of netted cloth and the rotated manually at a high speed. The pure honey thus extracted is collected through a basal outlet. The comb foundations remain undamaged and can be used further. The honey extractors may be of following three types :

a) Tangential (b) Radial (c) Parallel-radial extractor: All these work on the same principle, but tangential one is used by small scale bee keepers where as the other two are used by large scale bee keepers. Honey so obtained is filtered and stored for use.

viii) *Drone Trap:* The drone excluder on trap consists of a piece of wood slightly longer than the entrace of the hive with a shallow opening of 37 mm deep. It is used to get rid of the drones as they are unable to get back into the hive. Moreover its also used to catch swarm for setting new apiary.

Bee Flora

Bees feed on pollens of plants which is rich in proteins and for making honey they require nectara as raw material. Since, bee's donot visit all types of flowers, mostly those which donot smell well. Thus they too are confined to some selected flowers to obtain nectar. The plants that are used to obtain pollens and nectar collectively form bee flora. The timeperiod during which most of the plants flower, so that maximum nectar (via flowers) gets available to them is called honey flow-period. This period usually remains for 4-10 weeks of early spring and summer. Some of the important flower yielding plants include - litchi chinensis, Syzygium, Eucalyptus, Terminalia, Toona Eiliata, Tamarindus indica Brassicaceae plants etc. Thus bee keeping also includes cultivation of those plants which form frequently visiting sites of these insects. so that maximum yield is obtained.

Bee keeping Tools

- Smoker.
- Beekeeping Brush. Soft hair brush.
- Scraper/hive tool. You may use a hive tool/scraper for scraping wax, loosen hive parts and manipulate frames.
- Wax cutter or knife either normal or electrical.
- Wax heating pot.
- Wax sheet maker.
- Comb pulle.

Bee Colony and Social Behavior and Reproduction: A colony of honey bees consists of three kinds of individuals i.e., trimorphic, consisting a queen, drones and workers. Population of an average sized colony consists of one adult queen, about 100 drones and 60,000 workers. Drones and queen are concerned solely with reproduction where as workers perform other duties including production of royal jelly, wax, building hives, rearing larvae, cleaning hive, disposing-dead bees etc. Thus, they reflect highly organised division of labour.

i) **Worker Bee**: These are under developed females produced from the fertilized eggs but remain sterile due to non-availability of royal jelly for feeding during their development. Worker bees are smallest member of colony and make up the largest number of colony individuals. They are black or brownish in colour with body densely covered with hair. Like all other insects, the body of worker bee is divisible into three regions. Head, thorax and Abdomen.

1. **Head.** It is a wide triangular structure with the apex pointed below. It bears a pair of large compound eyes and three ocelli on the middle of its top. A pair of short and thirteen jointed antennae are borne on the middle of face and probably serve as tactile and gustatory organs. Mouth parts of honey bee are of chewing and lapping type, adapted for sucking nectar from flowers and moulding the wax. Mouth parts arise from the bottom of head.

2. **Thorax.** Usually thorax is divided into three segments, anterior prothorax, middle mesothorax and posterior metathorax. Each of these segments bear a pair of legs and a pair of wings is borne by each of the mesothorax as well as metathorax.

3. **Abdomen.** It consists of six visible segments and bears the wax glands and the sting. Wax glands lie on the ventral surface of the last 4 visible segments of abdomen; wax secreted is used for building the cells of the honey comb. Where as sting is the modified ovipositor used for injecting poison for protection. They are responsible for all the work necessary for the maintenance and welfare of colony, such as:

   i) Building of comb with wax.

   ii) Collection of honey, pollen and water for the use of colony.

   iii) Protection of colony.

   iv) Feeding of larvae and queen with royal jelly.

   v) Help to maintain the temperature of the colony as per the requirement of environment.

vi) Cleaning of comb and burial of dead bees.

vii) Attending queen etc.

i) **The worker bees** show a remarkable example of sacrifies, initially for the first three days, they are cleaners as they clean walls and floor of empty cells of the hive for the reuse. From days 3-10, they act as nurses, their pharyngeal gland become active and secrete royal Jelly for young larvae, queen, drones and those older larvae which are set to develope into future queen. After 10 days, their pharyngeal glands atrophise and wax glands develope and start functioning to produce wax used for making cells of the honey combs thus called builders-upto 16th day. From 16-20th day they receive pollen loads and place then in the comb into pollen cells. Between 20-23 days, they guard at the entrance of the hive defending the colony from enemies during which they pay their lives unpaid just to ensure protection, referred as guards. During the last few days of their life they explore new sources of food including nectar, pollen and water. They are called foragers and soon death arrives to ceasetheir service.

ii) **Queen:** Queen is the only fertile female in a colony and every colony has a queen from the eggs of which other castes of the colony develope, hence truely called mother of colony. She is enlongated, 15-20 mm long and is easily distinguished by her long tapering abdomen, short legs and wings. She has shorter mouth parts and pointed mendibles. She is without salivary glands and is unable to produce wax and honey or gather pollens or nectar. Queen arises from a fertilized egg and queen larva specially feeds on royal jelly. She has highest life span in all the castes and produces upto 15,00,000 eggs during her life time. A queen lives for 2-5 years sand when it's weak or unable to lay eggs, its replaced by one of the daughter queens.

iii) **Drones:** They represent male bees and can been seen flying idly near the hive in the sun-shine. They develope from unfertilized eggs and their number hardly exceeds 100 in a colony. They are of intermediate size (15-17 mm long), but considerably stouter and broader. They possess very large eyes, small pointed mendibles and lack wax producing glands, pollen collecting apparatus and a sting. They are unable to feed those seves and are solely dependent on workers for food. They die if not fed by workers. The sole function of this caste is to fertilize virgin female, when she leaves the colony on her nuptial flight. The drone follow in swarm and only one among them mates with the queen and dies immediately after the coupulation. They are driven out of the hive soon after the swarming season is over.

Life History

1. **Swarming**: Swarming is a process of leaving off the colony by the queen along with some members of other castes of the colony to relieve the over crowding and found new colony. Swarming takes place during early summers or in spring. During these days conditions are favourable and plenty of food is available. Basically when the aviability of food is much higher, the queen lays large number of eggs producing greater number of individual causing over crowding and bees start preparation for swarming. During this time at the bottom of the comb, queen daughter cells are build and when the new queen emerges out. Old queen leaves the hive i.e., (prime swarm) accompanied by a large number of old workers and drones. The swarm soon settles down, often on the branch of a tree to build and establish new hive swarming is always disliked by the keepers as it heavily reduces their yield.

2. **Nuptial flight**: It's also called marriage flight. About a week after emergence, the new queen takes her first aerial flight followed by a swarm of drones. The queen flies very high and the unfit drones gradually drop out of the race. The most strong and fit drone left in the race, mates with her. During mating the genital parts of male are forced out with such a great pressure that it dies in the course. The queen mates only once in her life. The sperms received by the queen are stored in spermatheca and are used to fertilize the egg as long as she lives. The pair - (dead drone and queen) falls to the ground and queen pulls herself away and returns to the hive permanently untill prime swarm.

3. **Oviposition:** After copulation the female takes some time and then starts laying eggs. As per the law of probability of nature fertilized and unfertilized eggs which are diploid and haploid respectively are laid. One egg is laid in each cell. However the fertilization is controlled by queen itself and fertilized eggs are laid in worker or queen cells and unfertilized in drone cells. About 1,500 eggs are laid by the queen in a day, producing 15, 00,000 eggs in her life time.

4. **Grub:** After 3-4 days, small larvae or grubs hatch out from the eggs. From the fertilized eggs, queen and workers develope and from the unfertilized eggs drones develope parthenogenetically. The grubs are cylindrical and light yellow in colour and have no legs or eyes.

The formation of a queen or worker depends on the diet on which the larvae are fed. For the first two-three days all type of larvae are fed on royal jelly, there after the drones and workers are fed on bee bread (mixture of honey and pollens) and developing queens are contineously fed on Royal Jelly. The grub period lasts for 5-6 days.

5. **Pupa**: A pupa is enclosed in a silken cocoon secreted by the larva. The pupal period lasts for 7-14 day depending upon the type of adult to be produced. It takes the queen on an average 13 days, the workers 18 days and the drones 21 days to complete metamorphosis and emerge out as adults

Caste	Egg	Larva	Pupa	Total
Queen	3 days	6–6.5 days	7–7.5 days	15–17 days
Worker	3 days	6–8 days	10–11 days	18–28 days
Drone	3 days	6.5–10 days	11–15 days	21–24 days

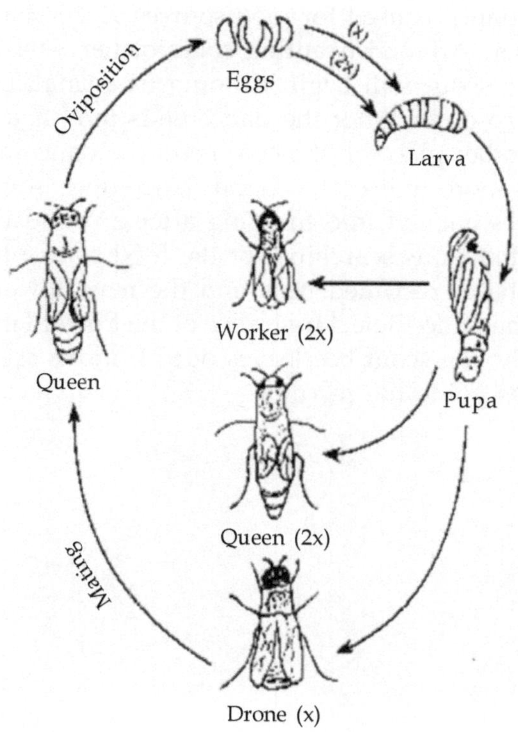

Social behavior in bees has a number of advantages. One of the most important of these is the ability to quickly mobilize a large number of foragers to gather floral resources that may only be available for a short period of time. The ability to communicate location with such precision is one of the most interesting behaviors of a very interesting insect. The recruitment of foragers from a hive begins when a scout bee returns to the hive engorged with nectar from a newly found nectar source. She begins by spending 30-45 seconds regurgitating and distributing nectar to bees waiting in the hive. Once her generosity has garnered an audience, the dancing begins. There are 2 types of bee dances: the round dance and the tail-wagging or waggle dance, with a transitional form known as the sickle dance. In all cases the quality and quantity of the food source determines the liveliness of the dances. If the nectar source is of excellent quality, nearly all foragers will dance enthusiastically and at length each time they return from foraging. Food sources of lower quality will produce fewer, shorter, and less vigorous dances; recruiting fewer new foragers.

The Round Dance

The round dance is used for food sources 25-100 meters away from the hive or closer. After distributing some of her new-found nectar to waiting bees the scout will begin running in a small circle, switching direction every so often. After the dance ends food is again distributed at this or some other place on the comb and the dance may be repeated three or (rarely) more times. The round dance does not give directional information. Bees elicited into foraging after a round dance fly out of the hive in all directions searching for the food source they know must be there. Odor helps recruited bees find the new flowers in two ways. Bees watching the dance detect fragrance of the flower left on the dancing bee. Additionally, the scout bee leaves odor from its scent gland on the flower that helps guide the recruits.

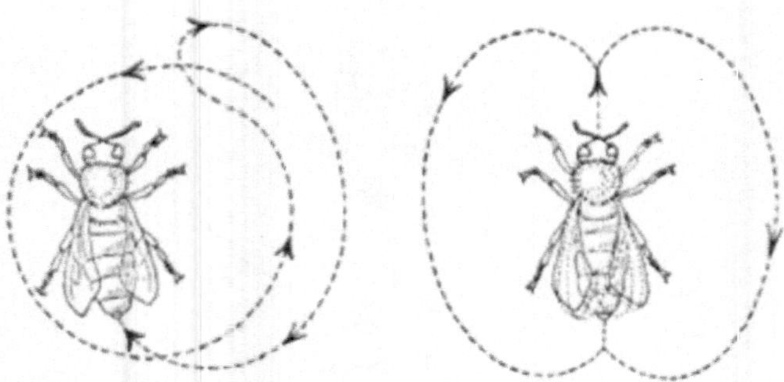

Fig. 1. Round dance **Fig. 2.** Waggle dance

The Round dance and Waggle dance, as described by Karl Von Frisch, 1976

The Waggle Dance of Honey bees

As the food source becomes more distant the round dance is replaced by the waggle dance. There is a gradual transition between the round and waggle dance, taking place through either a figure two or sickle shaped pattern shown here is a dancing bee on a swarm. The marked bee has returned from a sugar water feeder. She is seen here being unloaded by a number of bees, and then begins to dance, which communicates the source of sugar water related to the current sun azimuth. This recording was produced at the University of California, Riverside by Kirk Visscher. The waggle dance includes information about the direction and energy required to fly to the goal. Energy expenditure (or distance) is indicated by the length of time it takes to make one circuit. For example a bee may dance 8-9 circuits in 15 seconds for a food source 200 meters away, 4-5 for a food source 1000 meters away, and 3 circuits in 15 seconds for a food source 2000 meters away. Direction of the food source is indicated by the direction the dancer faces during the straight portion of the dance when the bee is waggling. If she waggles while facing straight upward, than the food source may be found in the direction of the sunIf she waggles at an angle 60 degrees to the left of upward the food source may be found 60 degrees to the left of the sun. Similarly, if the dancer waggles 120 degrees to the right of upward, the food source may be found 210 degrees to the right of the sun. The dancer emits sounds during the waggle run that help the recruits determine direction in the darkness of the hive.

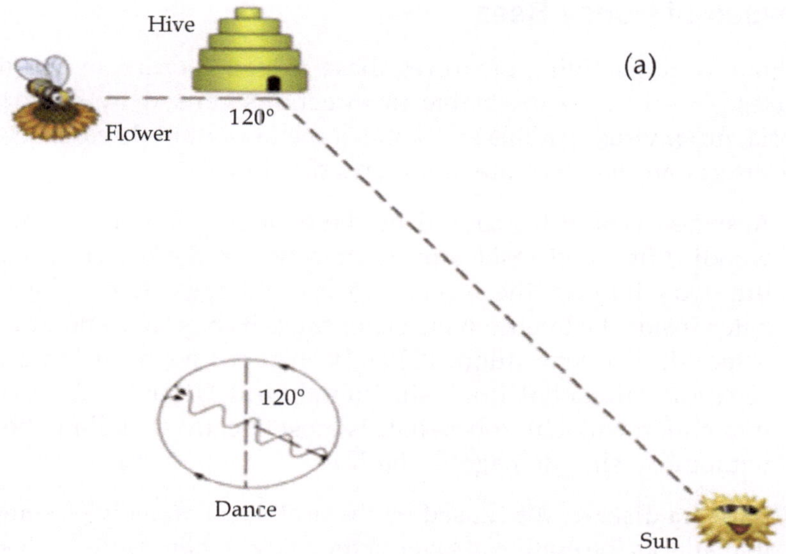

Photo: Bill Tietjen, Bellarmine University

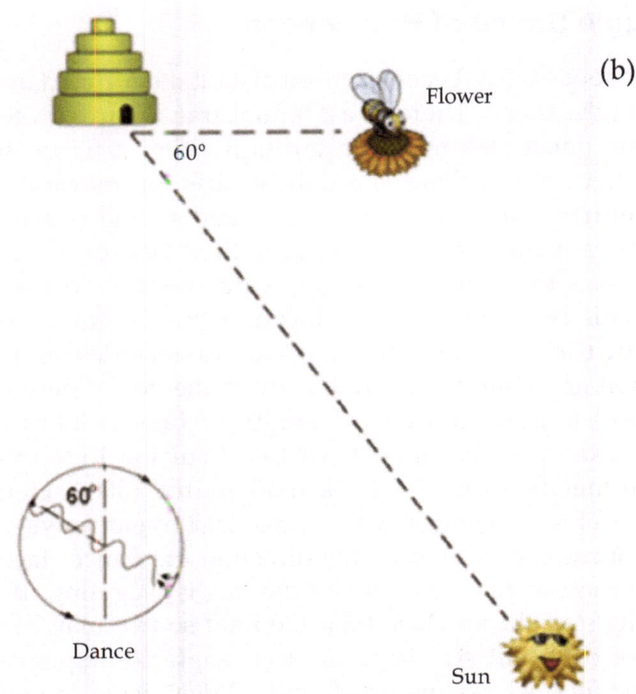

Photo: Bill Tietjen, Bellarmine University

(Source: The information in this article was taken from "The Dance Language and Orientation of Bees" by Karl von Frisch).

Diseases of Honey Bees

Like all other living creatures these insects too are susceptible to diseases. Mostly they are liable to infections caused by protozoans, bacteria, fungi viruses, mites etc. Some of the important diseases to which these insects are liable of are summerised below:

1. **Acarine disease**: It's caused by the endoparasite namely Acarapis woodi (Mite) that resides in the treachea of the insect. It feeds on the body fluid of the insect and lays its eggs there. These eggs hatch inside the trachea which later cause its blockage and ultimately insect dies of starvation. It has been reported from Jammu and Kashmir Himachal Pradesh; Punjab and Uttar Pradesh Smoke fumigation with chlorobeuzilate is most effective in killing the mites without causing damage to the bees.

2. **Nosema disease**: It's caused by the protozoan, namely Nosema apis that attacks the epithelial layer of the midgut that results in dysentry. Since, the faeces of the infected bees contain spores of the protozoan

that is consumed by the cleaner bees, hence they also get affected. During the infection stomach of the suffer or becomes swollen and white in colour. This disease spread fast during winter and spring. Its most common in temperate countries like U.S.A. etc. Fumagillin – a new drug, in concentrated syrup considerably counters nosema disease. Moreover providing running water and full sunlight also reduces its effects. Sterilization of brood boxes and frame hives with 40% of formaline fumes also minimize the attack.

3. **Amoebic disease:** This disease too is caused by protozoan namely. Malphighamoeba mellificae. It attacks the malphighian tubules and midgut epithelium and are excreted along with faecal matter, that is consumed by the cleaner bees which become the victims later. To curb this disease, proper hygenic conditions should be maintained and all the equipments used in the apiary must be sterilized.

4. **Septicemia:** This is one among the various bacterial diseases caused by Bacillus apiscepticus. It affects the larval stage of the insect and symptoms include sunken and perforated cappings of sealed brood. The dead larva produces foul smell. This disease spreads through contaminated water, keeping conditions hygenic and instruments sterilized helps to minimize its affect.

5. **Stone brood disease**: A fungal disease caused by fungus *Aspergillus flavus*, the spore of which sprouts with in the alimentary canal of the insect larva. Sometimes it also infests adults. The insects after death turn into hard stone like mummies hence got its name as stone brood disease.

6. **Sacbrood disease**: Its one of the earliest disease to which these insects have been susceptible, it's caused by a retrovirus (RNA. genome) belonging to family picornaviridae. It infests the larvae of honey bees and they fail to pupate. The dead larva appears sac like creature of dark brown colour.

7. **Acute bee paralysis**: This disease is caused by Acute Bee Paralysis Virus (ABPV), it affects the adults and is responsible for their high death rate. Infested bees show partial hairlessness on body, trembling, sprawled legs and wings. Death occurs in a few days after infestation and body of sufferer is completely paralysed. Recently Indian bees have become susceptible to viral infections and is most important problem of bee keeping in India. Maintaining proper hygenic conditions, sterilization of tools incorporated in apiary, weekly visits to check the conditions of hives, formaline spraying etc. helps in checking these organisms from diseases.

Bee Enemies

Some of the important enemies of bees are summarised below:

1. **Wax Moths:** Greater wax moth *(Galleria mellonela)* and lesser wax moth *(Achroia grisella)* are two species of wax moths that damage the apiary and consequently reduce our yield. The female moth lays eggs inside the hive on the covered portion of the combs, in cracks, and on the bottom board of the hive. The caterpillars emerged tunnel the combs and invade them with webs and develope into pupae. They consume stored pollens and bees wax. If this pest is not checked in time, the whole combs may be destroyed and bees may abscond (leaving the hive). Keeping apiary hygenic, sterilization of tools used, periodic fumigation helps to check their attack.
2. **Ants**: Some of the species of Ants that attack apiaries include *Dorylus labiatus, Componotus compressus, Monomorium indicum* and *M. destructor.* They generally attack weak colonies and are trouble some after rain. They take away honey Pollens and Brood. Destroying ant nests with fire in the vicinity of bee hives helps us to check their attack, though its not an efficient method to curb them. Most importantly the apiaries must be placed on stands whose legs should be dipped in water cups to check their attack.
3. **Wasps:** Gaint wasps, *Vespa orientalis, Vespa magnifica* etc. are some of the species of vespa that cause serious attack to bees. They catch bees and eat them in their own nests. They are attack is at peak during rainly seasons and remain dormant during winter; their nests must either be burned or be sprayed by insecticides to protect the colonies.
4. **Birds:** The green bea eater *(Merops orientalis)* and the kings crow *(Dicrurus macrocercus)* are two important birds that catch bees during their flight and eat them. Since the honey bees carrying nectar and pollens are sluggish in flying, Birds easily catch them; hence colonial population and honey yield is highly reduced. Acetyl guns should be used to scare birds.

Frogs and Lizards: Frogs and lizards are often seen near the hives and cause damage to the honey bees. Nocturnal lizards often kill bees which have lost their way.

Honey Bee Products

Honey: This is the most important product of Apiculture. It is prepared by worker honey bees from the nectar collected by them from plants (flowers). The bees draw nectar via proboscis and carry it in their crops. When they reach in the hive, they regurgitate and its stored in cells meant for its storage. There it is subjected to enzymatic actions and finally ripes into honey-viscous sweet liquid. Honey is believed to be highly nutritious. It has been estimated that 2.1 gm of honey contain 67 kilo calories of energy. Thus 200 gms of honey provide some energy as 330 gm of meat or 11.5 ltr of milk or 1.6 kg of cream. However it is important to note that the colour, aroma and constitution may vary, like in case of apple honey *(Pyrus malus)*, colour of honey is pale yellow with 31.6% glucose where as in case of Acacia honey (Robina Pseudoaccacia – Nectariferous plant) it is of transparent look and contains about 38% of glucose. Thus there is exists close correlation between colour, aroma, constitution and nectariferous plant (plant from which nectar is obtained). It is well established that bees also prepare honey by using wild poisonous plants like. Kalmia, Rhododendron ledum etc. as nectariferous plants. Such honey is toxic and a dose of 14 gm of such honey may kill a pig.

S.No.	Constituents	Quantity
1	Water	17.2%
2	Fructose	38.2%
3	Glucose	31.3%
4	Sucrose	1.3%
5	Maltose	7.3%
6	Oligo saccharides and Polly saccharides	1.5%
7	Gluconicacid	0.57%
8	Minerals	0.17%
9	Proteins and Amino acids	0.04%
10	Enzymes vitamins, aromatic substancesand other constituents	0.03%

Uses of Honey

i) **Honey as a Food:** Honey is a high energy carbohydrate food. it is the best source of heat and energy giving over 5,500 calories/kg. Our body system readily absorbs sugar, mineral, vitamins and other elements of honey. Honey is useful to a healthy and sick person as well, its taken by any age being even the newly born babies could be fed. Mostly honey is taken with bread milk, cereals, fruits etc. Its used in preperation of jams, jellies beverages, cakes, pastries, chocolates etc. It can be taken in its natural form too or even with water or milk.

ii) **Honey as Medicine:** Honey has its great medicinal value. It's generally used in Ayurvedic and Unani systems of medicine treatment. It's used as a laxative, a blood purifier against cold, fever, cough etc. Most importantly linden honey is used against colds; eucalyptus honey removes nervous disorder weakness, mental fatigue, tuberculosis etc. Medically some of the important uses of honey include:

a) *Gastric medicine:* Its traditionally used to cure gastric ulcers as it provides relief from hyperacidity.

b) *Cardiac medicine:* Honey is also used against heart problems, it causes expension of veins and improves circulation through coronary arteries. 50-80 gm of daily dose of honey improves heart functioning and is often recommended to heart patients.

c) *Antibiotic uses:* It has important antibiotic properties and is therefore used to treat various bacterial and fungal infections.

d) *Neuro medicine:* Its used to treat headache. Some honey types contain measurable amounts of dopamine seratonin etc. and are recomended to psychataric patients.

e) *Eye medicine:* Honey is also used to cure eye burns, eyelid swellings, conjuctiva cornea swelling, soreson corneal, membrane etc. Besides these uses, its also used to cure various types of wounds (necrotic), ulcers, bacteria infections like TB, helps in increasing Hb% etc.

iii) **As a Cosmetics:** Honey is a quick killer of germs, therefore its incorporated in cosmetic creams to protect skin from germ attacks honey also penetrates the skin and supplies glucose to the muscle layer of the skin that helps it to glow. Due to its natural antiseptic property its heavily used in creams, face washes, lotions, sun guards etc.

iv) **Religious uses:** Honey is considered as a food of heaven and its use is recommended in various religious book like the holy Quran etc. Many festivals require honey in various rites to please gods that is why its sometimes called food of gods in Hinduism. Besides these values, honey is used in distillaries for making alcoholic drinks and in poultary and fishing industries. In laboratories its used as growth stimulant of plants, for the bacterial culture etc. More importantly it forms the diet of several insects and is used in making poison baits to save many fruits from fruit flies.

Propolis: Sap or resin gathered by bees from the bark and leaves of trees is mixed up with nectar and enzymes to form propolis. It is used to repair their hives. Most importantly its also used to protect hive from external contaminations and unpleasent odour. It contain almost all types of vitamins excepts vitamin K and has been used for wound healing and tissue regeration. Propolis is an excellent natural antibiotic and immune system booster. Due to its high anti biotic potential bees use propolis at the entrance of their hive so that they get sterilized before entering the hive. Its useful in allergies, burns, sore throats, nasal congestion, respiratory infections, skin disorders etc.

Beeswax: Beeswax is secreted by the worker honey bees via specialised epidermal glands present on the underside of the abdomen refered as wax glands, which are of 4 pairs. Its yellowish white solid waxy material containing a mixture of cerotic acids myricle palmitate. Its used for constructing hive and its estimated that honey bees consume 8-16 kg of honey to produce 1 kg of beeswax. Commercial bees wax available in the market is obtained from the wild hives of apis dorsata. Its widely used as a thick base of creams, lipsticks, beauty lotions etc. Its estimated that more than 75% of total worldwide production of beeswax enters in cosmetic and pharmaceutical industries. Medically it acts as a base for various ointments and forms coating of various tablets. Other uses include manufacturing of candles, polishes, chewing gurms, adhesives etc. Its of common use in the microtomy work in laboratories to prepare blocks of tissue where its mixed with common wax.

Royal Jelly: 'Crown Jewel of beehive' is the secretion of hypopharyngeal glands of worker bees when these glands are active i.e., between 3-10 days of age. All type of larvae is initially fed on royal jelly, but it is only the developing queen that feeds on royal jelly throughout. It's a rich source of proteins, minerals, anti-oxidents, vitamins, fats and other substances. It's of great economic importance and has been used against several bacterial and fungal infections. It is believed to be a general tonic, anticold, increases potency and delays aging in humans. World production of royal jelly is about 500 tonnes and is marketed in capsules creams, soft gels, ointments etc.

Bee Venom: Its secreted by poison glands of worker bees and its medicinal use form apitherapy. It contains melittin, apamine, histamine, tryptophan, phospholipase, hyaluronidase, seratonin, dopamine etc. Bee venom is useful to lower blood pressure and cholesterol level it is used against inflammation and conditions like rheumatoid and osteoarthrits. It is used for neautralizing alcohol poisoning and relieve pain due to its antidote and analgesic properties respectively.

Pollinators: Pollination refers to the transfer of pollens from anther of stamens to the stigma of carpet, to ensure fertilization. The agencies which carry these pollen grains are refered as pollinating agents. Foraging (Field bees) bees have to visit from one flower to other, due to their sticky body and legs pollen grains often get attached with them and accordingly they get transfered from one flower to another flower. Thus, they (bees) facilitate cross fertilization to produce better yield.

Selected References

Askey R.R. (1971). Parasitic Insect, H.E. Books, London.

Chaponan R.F. (1982). The Insect Structure and Function. 3rd Edition, Harvard Publishing Press, Combridge MA.

Elzinga, R.J. (1978). Fundamental of Entomology. Prentice-Hall of India Pvt. Ltd. New Delhi.

Ghorai N. (1995). Lac-culture in India. International Books and Periodicals Supply Service, New Delhi.

Grimaldi D. and Engel M.S. (2005). Evolution of the Insects. Cambridge Univ. Press, New York.

Kumar and Dhiman (2014). Biology, Ecology and Population Dynamics of *C. stolli* Wolf (Penta). Publisher: Scholar's Press and LAP Publishing Company.

Lagreca, M. (1980). Origin and Evolution of Wings and Flight in Insects. Bull. Zool. 47: 65-82.

Mani, M.S. (1989). Indian Insects, Satish Book Enterprise, Agra.

Mathur and Upadhyay. A Text Book of Entomology. Aman Publishing House, Meerut (India).

Nigam, P.M. (1989). Text Book of Agricultural Entomology. Emkay Publication, New Delhi.

Plak Prakash (2009). Laboratory Manual of Entomology Published by New Age International (P) Ltd., Publishers 4835/24 Ansari Road Daryagang, New Delhi 110002.

Pruthi, H.S. (1969). Text Book on Agricultural Entomology, ICAR, New Delhi.

Sehgal, P.K. (2014-17). Fundamental of Agricultural Entomology, Kalyani Publishers, New Delhi, pp. 1-401.

Singh Rajendra (2012). The Elements of Entomology. First Edition. Rastogi Publications, Shivaji Road, Meenit (India).

Singh S. Bee Keeping in India ICAR, New Delhi (1975).

Srivastava, K.P. (1993). Text Book of Applied Entomology, Vol I and II. Kalyani Publishers, Ludhiana, India.

Tembhare B.D. (2012). Modern Entomology. The Publishing House Pvt. Ltd., Delhi.

Winston M.L. (1987). The Biology of Honey Bee. Cambridge Univ. Press.

645

B.Sc. (Agriculture) (Second Semester)

EXAMINATION, 2017

Paper Sixth

INTRODUCTORY ENTOMOLOGY

Time: 3 Hours **Maximum Marks:** 35

Note: Attempt all the Sections as directed.

Section—A

(Objective Type Questions)

Note : Attempt all questions. Each question carries 1 mark.

1. How many segments are found in insect antenna?

   a) Three b) Two

   c) Five d) Ten

2. Natatorial type of leg is modified for :

   a) Walking b) Jumping

   c) Swimming d) Digging

3. Which of the following is a pterothorax ?

   a) Prothorax b) Mesothorax

   c) Metathorax d) Both (b) and (c)

4. Gnathocephalonic segment possesses the following appendages:

   a) Mandibles b) Maxillae

   c) Labium d) All of these

Fill in the blanks :

5. Two pairs of antennae are present in ....

6. Hind leg of grasshopper is modified for ....

7. Jonhston organ present on which antennal segment......
8. Antennae are absent in ....

Write true / false :

9. Insect body is divided in four segments.
10. Siphoning type of mouthpart is found in moth.

Section—B

(Short Answer Type Questions)

Note : Attempt any five questions. Each question carries 2 mark.

11. Who is author of "Indian Insect Pests" ?
12. What are the major functions of insect antenna ?
13. How many types of insect head ? Describe each with suitable diagram and example.
14. What is the difference between simple and compound eye ?
15. What are endocrine glands ? Describe their role in insect growth and development.
16. Differentiate between pterygota and apterygota.
17. Define entomology and write main characters of class hexapoda.

Section—C

(Long Answer Type Questions)

Note : Attempt any three questions. Each question carries 5 mark.

18. Classify the insect mouthparts on the basis of type of food. Describe generalized type of insect mouthpart with well labeled diagram.
19. Describe the digestive system of insect and process of digestion with well labeled diagram.
20. Give a detailed account of insect reproductive system with well labeled diagrams.
21. Name various modifications of insect legs and describe each with diagrams and examples.

22. Write short notes on any two of the following with suitable diagram and example :

    i) Metamorphosis

    ii) Nervous system

    iii) Characters of order diptera

    iv) Characters of order coleoptera

    v) Filter chamber

    vi) Body segmentation

    vii) Function and types of insects wing

    viii) Insect dominance

605

B.Sc. (4 Years) (Fifth Semester)

EXAMINATION, 2016-2017

Paper Seventh

HORTICULTURE

(Insect Pests of Fruit Plantation, Medicinal and Aromatic Crops)

Time: 3 Hours **Maximum Marks:** 35

Note: This question paper consists of three Sections.

Section—A

(Objective Type Questions)

Note : Attempt all questions. Each question carries 1 mark.

Fill in the blanks :

1. Bactrocera dorsalis hibernate in ....
2. Sanvose scale was introduced in India (Kashmir) from France in year ....
3. The dead body of female mango mealy bug is found sticking to their eggs. (True/False)
4. Collection and distribution of egg, larvae, pupa and adults are chemical pest management. (True/False)
5. Fruit fly lay eggs in soil. (True/False)
6. Correct identification of insect-pest is basic rule of pest management. (True/False)
7. Scientific name of Grey weevil is Myllocerotis subfaciatus. (True/False)
8. Tent caterpillar is a pest of mango. (True/False)

Choose the correct answer.

9. Sun drying is a method of IPM :

a) Chemical b) Physical

c) Mechanical

10. Mango Hopper is a :

a) Polyphagous pest b) Monophagous pest

c) Oligophagous pest

Section—B

(Short Answer Type Questions)

Note : Attempt any five questions. Each question carries 2 mark.

11. Write nature of damage and management of Black Headed caterpillar.
12. How to manage mint aphid ?
13. What is the difference between control and management ?
14. How save citrus orchard from citrus psylla infestion ?
15. Write common name, scientific name, family, order and damaging stage of any five borer insect pests of fruit plantation crops.
16. Who transmits Bunchy top disease of Banana ?
17. Write nature of damage and management of citrus whitefly.

Section—C

(Long Answer Type Questions)

Note : Attempt any three questions. Each question carries 5 mark.

18. Write nature of damage, damage stage and management of my five of following insects :

i) Sanvose scale ii) Peach leaf curl aphid

iii) Peach stem borer iv) Bark eating caterpillar

v) Termites vi) White grub

vii) Mango stem borer

19. Write common name, scientific name, order, family and damaging stage of the major insect-pest of citrus. Describe nature of damage and management of citrus butter fly, citrus aphid and fruit sucking moth.

20. Enlist major insect pest of Banana. Describe three of them with their distribution, nature of damage and management.

21. Write short notes on any five of the following :

    a) Pesticides residue

    b) EIL

    c) MRL (maximum residue limits)

    d) Botanicals in pest management

    e) Organic pest management

    f) Integrated pest management for stored fruit

    g) Mechanical method of pest management

22. Define the term monitoring, survey, surveillance and forecasting and explain their role in integrated pest management.

582

B.Sc. (Part II) (4 Years) (Third Semester)

EXAMINATION, 2015-2016

Paper Second

HORTICULTURE

(Fundamentals of Entomology and Nematology)

Time: 3 Hours **Maximum Marks:** 35

Note: This question paper consists is divided into three Sections.

Section—A

(Objective Type Questions)

Note : Attempt all questions. Each question carries 1 mark.

1. In what order of insect would you find grasshoppers ?
2. Body of Hexapoda class is divided into .... parts.
3. A mite belongs to the class ....
4. Arthropoda is the largest group of insects. (True/False)
5. Damaging stages of Coleoptera order are ....
6. Termites damages plants mostly during at night hours. (True/False)
7. Blood circulatory organs are found in nematodes. (True/False)
8. Body of nematode is covered by ....
9. White grub is a polyphagous insect. (True/False)
10. .... and .... are the damaging stages of painted bug of mustard.

Section—B

(Short Answer Type Questions)

Note : Attempt any five questions. Each question carries 2 mark.

11. Write a short note on pest.

12. Describe the class Hexapoda with suitable example.
13. How aphids damage the fruit crops ? What control measures are suggested by you to control the aphid infestation ?
14. Differentiate between economic injury level and economic threshold level.
15. Describe about the insects.
16. Describe the mouthparts of mites.
17. Write about the direct metamorphosis.

Section—C

(Long Answer Type Questions)

Note : Attempt any three questions. Each question carries 5 mark.

18. Give a detailed account on modification of insect mouths with suitable labelled diagrams.
19. Focus on the importance and scope of integrated insect pest management.
20. Describe insect digestive and excretory system with suitable labelled figures.
21. Describe different external body parts of locust with well labelled diagrams.
22. Write short notes on the following :
    i) Symptoms developed and control measures of nematodes
    ii) Types of insect antennae.

659

B.Sc. (Agriculture) (Fourth Semester)

EXAMINATION, 2015

Paper Fifth

INSECT PESTS AND THEIR MANAGEMENT

Time: 3 Hours **Maximum Marks:** 35

Note: This question paper consists is divided into three Sections.

खण्ड—अ

Section—A

वस्तुनिष्ठ प्रश्न

(Objective Type Questions)

नोटः सभी प्रश्नों के उत्तर दीजिए। प्रत्येक प्रश्न 1 अंक का है।

Note : Attempt all questions. Each question carries 1 mark.

1. सरसों को बंदगोभी में .... के प्रबंधन के लिए ट्रैप फसल के रूप में उगाया जाता है।

   Mustard is grown as a trap crop for the management of .... in cabbage.

2. भारत में सब्जियों पर .... कीटनाशक का प्रयोग प्रतिबंधित है।

   Insecticide restricted for use on vegetables in India is ....

3. फैरोमोन ट्रैप .... पतंगे को आकर्षित करने के लिए प्रयोग किया जाता है।

   Pheromone trap is used to attract .... moth.

4. एकलभक्षी कीट का उदाहरण .... है।

   An example of monophagous insect is ....

5. धान का पीला तना भेदक .... पर अण्डे देता है।

   Yellow steam borer of rice lays eggs on ....

6. कुरमुला कीट का वयस्क पौधे की जड़ों पर खाता है। (सत्य/असत्य)

   The adult white grub feeds on roots of plant. (True/False)

7. वूली एफिड अमरूद का नाशी कीट है। (सत्य/असत्य)

   Lady bird beetle is parasitoid of aphids. (True/False)

8. तम्बाकू की सूंडी एकल अण्डे देती है। (सत्य/असत्य)

   Tobacco caterpillar lays eggs singly. (True/False)

9. लेडी बर्ड बीटल माहू कीट का पैरासिटायड है। (सत्य/असत्य)

   Lady bird beetle is parasitoid of aphids. (True/False)

10. धान में हापर बर्न का कारण भूरा तना फुदका है। (सत्य/असत्य)

    Hopper burn in ripe is caused by Brown plant hopper. (True/False)

खण्ड–ब

Section—B

लघु उत्तरीय प्रश्न

(Short Answer Type Questions)

नोटः किन्हीं पाँच प्रश्नों के उत्तर दीजिए। प्रत्येक प्रश्न 2 अंकों का है।

Note : Attempt all five questions. Each question carries 2 mark.

11. निम्नलिखित कीटों का वैज्ञानिक नाम एवं कुल लिखिएः

    अ) मक्का प्ररोह मक्खी   ब) सरसों का माहू

    स) आम का फुदका   द) तम्बाकू की सूंडी

    Write scientific name and family of the following insects :

    a) Maize shoot fly   b) Mustard aphid

    c) Mango hopper   d) Tobacco caterpillar

12. निम्नलिखित कीटों की अण्डे देने एवं प्यूपेशन की जगह लिखिएः

    अ) बैंगन का प्ररोह एवं फल भेदक   ब) सरसों की आरा मक्खी

    स) गन्ने का जड़ भेदक   द) चने का कटुआ कीट

Write egg laying and pupation site of the following insects :

a) Brinjal shoot and fruit borer b) Mustard saw fly

c) Sugarcane root borer d) Gram cut worm

13. निम्नलिखित कीटों द्वारा किए गए क्षति के लक्षण दीजिए:

अ) प्याज थ्रिप्स ब) नींबू की सूडी

स) धान की गाल मक्खी द) दीमक

Give damage symptoms of the following insects :

a) Onion thrips b) Citrus caterpillar

c) Paddy gall fly d) Termite

14. एक कीटनाशी जीव कैसे बन जाता है? उदाहरण सहित उत्तर दीजिए।

How does an insect become pest ? Answer with suitable examples

15. रासायनिक कीटनाशकों का सब्जियों पर सावधानीपूर्वक प्रयोग क्यों करना चाहिए? विस्तारित कीजिए।

Why should chemical insecticides be used in vegetable with due care ? Elaborate.

16. निम्नलिखित के लिए उपयुक्त पैस्टीसाइड अनुमोदित कीजिए:

अ) आम का मीली बग ब) गुलाबी बौड़ी शलभ

स) रैड स्पाइडर माइट द) चूहे

Recommend suitable pesticide for the following :

a) Mango mealy bug b) Pink boll worm

c) Red spider mite d) Rodents

17. रासायनिक कीटानाशकों के प्रयोग के फायदे एवं नुकसान लिखिए।

Write the merits and demerits in use of chemical insecticides.

खण्ड–स

Section—C

दीर्घ उत्तरीय प्रश्न

(Long Answer Type Questions)

नोटः किन्हीं तीन प्रश्नों के उत्तर दीजिए। प्रत्येक प्रश्न 5 अंकों का है।

Note : Attempt any three questions. Each question carries 5 mark.

18. कुरमुला कीट के पोषक पौधे, पहचान एवं जीवन–चक्र का वर्णन कीजिए।

    Describe Host range, identification and life cycle of white grub.

19. टमाटर के फल भेदक का क्षति का प्रकार एवं एकीकृत नाशीजीव प्रबन्धन लिखिए।

    Write nature of damage and IPM of tomato fruit borer.

20. कपास की विभिन्न कीटों की सारणी बनाइए एवं इनका एकीकृत नाशीजीव प्रबन्धन दीजिए।

    Enlist different insect pests of cotton and suggest their IPM.

21. नाशीजीव प्रबन्धन की नवीन तकनीकों का विस्तारपूर्वक वर्णन कीजिए।

    Give a detailed account of recent techniques in insect pest management.

22. आम के विभिन्न कीटों की सारणी बनाइए एवं इनका पर्यावरण हितैषी प्रबन्धन दीजिए।

    Enlist different insect pests of mango. Suggest their ecofriendly management.